Ecology *and the* Environment

Templeton Science and Religion Series

In our fast-paced and high-tech era, when visual information seems so dominant, the need for short and compelling books has increased. This conciseness and convenience is the goal of the Templeton Science and Religion Series. We have commissioned scientists in a range of fields to distill their experience and knowledge into a brief tour of their specialties. They are writing for a general audience, readers with interests in the sciences or the humanities, which includes religion and theology. The relationship between science and religion has been likened to four types of doorways. The first two enter a realm of "conflict" or "separation" between these two views of life and the world. The next two doorways, however, open to a world of "interaction" or "harmony" between science and religion. We have asked our authors to enter these latter doorways to judge the possibilities. They begin with their sciences and, in aiming to address religion, return with a wide variety of critical viewpoints. We hope these short books open intellectual doors of every kind to readers of all backgrounds.

Series Editors: J. Wentzel van Huyssteen & Khalil Chamcham
Project Editor: Larry Witham

Ecology
and the Environment

THE MECHANISMS, MARRING,
AND MAINTENANCE OF NATURE

R. J. Berry

TEMPLETON PRESS

Templeton Press
300 Conshohocken State Road, Suite 550
West Conshohocken, PA 19428
www.templetonpress.org

Typeset and designed by Gopa and Ted2, Inc.

Library of Congress Cataloging-in-Publication Data

Berry, R. J., 1934-
  Ecology and the environment : the mechanisms, marring, and maintenance of nature / R. J. Berry.
      p. cm.
  Includes bibliographical references and index.
   ISBN 978-1-59947-252-2 (pbk. : alk. paper) 1. Ecology.
  2. Environmentalism. I. Title.
  QH541.B46 2011
  577—dc23
                                                    2011018951

Printed in the United States of America

11 12 13 14 15 16   10 9 8 7 6 5 4 3 2 1

Contents

 Preface

I do not know what I may appear to the world; but to myself
I seem only to have been like a boy playing on the seashore,
and diverting myself in now and then finding a smoother pebble
or a prettier shell than ordinary, while the great ocean of truth lay
all discovered before me. . . . If I have seen a little further it is by
standing on the shoulders of Giants. —Isaac Newton

SIR PAUL NURSE, Nobel Laureate, president of the Royal Society of London, and former president of Rockefeller University, has declared himself unashamedly optimistic about the future. He believes that scientific achievements in the twentieth century have provided a realization that there is a true unity to life and that by the end of the present century, we will really know how life works. Crucially, he sees this coming by putting the molecular and genetic revolutions of the twentieth century into the context of three Es—evolution, ecology, and ethology, and quickening them by developing a greater focus on "great ideas."

This book is about some of these great ideas, together with lesser ones that have turned out to be mistaken, but were worthwhile because they stimulated thought and research. Although formally about only one of Nurse's Es, this short volume on ecology overlaps with his other two, evolution (the study of change over time) and ethology (the study of animal behavior). It aims to provide a window into the emerging shape of biology more widely and with it (in Nurse's words), the "great potential for understanding life and how it works with huge benefits for humanity, improving health, wealth creation and improving the quality of life."[1]

I come to the topics in this book from a long professional connection with the subject. I was first enthused about the natural world as a small boy through BBC broadcasts. My professional involvement began when I worked with Bernard Kettlewell on melanic moths (see p. 31), which led me to becoming an ecological geneticist—albeit working mainly on mice not moths. I have enjoyed and been excited by fieldwork in many parts of the world—on islands around Britain, and in Peru, the Antarctic, Hawaii, and the Marshall Islands. A major event for me was in 1970 when, in preparation for the first U.N. Conference on the Environment in Stockholm, a Christian publisher asked me to write a short book about the Christian approach to the environment. It drove me to read what philosophers and ethicists had written about the natural world, what the Bible said about creation, and what we as humans ought to be doing. Forty years on, it is good to report herein on my continuing efforts to link my scientific studies with my understanding as a Christian. I believe we all need to know more about the world on which we depend, and crucially, to ask ourselves how we ought to treat it. We need to become environmentally literate.

Commenting on the award of the 2011 Templeton Prize to Martin Rees (Isaac Newton's successor as president of the Royal Society, U.K. Astronomer Royal and Master of Trinity College, Cambridge), David Ford, Regius Professor of Divinity, University of Cambridge, wrote in the April 23, 2001, edition of the (London) *Times*: "[The award] has not only been encouraging to all those who see it as false and dangerous to suggest that science is necessarily opposed to religion; it has also underlined the importance for the coming century of alliances between wise science and wise faith (which may in some people go together). Perhaps a catalyst for this will be the joint commitment to ecology, biodiversity, conservation and a sustainable environment." My hope for this book is that it will encourage the forging of such links between "wise science" and "wise faith."

Ecology *and the* Environment

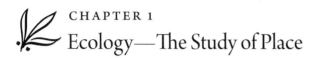

CHAPTER 1

Ecology—The Study of Place

The world is mud-luscious and puddle-wonderful.
e. e. cummings

LONG BEFORE the term "ecology" was invented, scientists involved themselves with the challenge of identifying and cataloging the vast number of life forms on earth. The British scientist J. B. S. Haldane reputedly, but probably apocryphally, said that the only thing we can know for certain about God is that he had an inordinate fondness for beetles. We don't know how many beetle species there are, but beetles are undoubtedly among the commonest creatures on the planet. If we put beetle species at a probable third of a million, then out of a total of eight to ten million species of all sorts of animals alive today, about one in twenty is a beetle.

Counting species in this sort of way does not mean very much. Some species are very rare (pandas and condors), others (rats, humans, and brown seaweed) occur throughout the globe. Some species eat others (lions and sharks), others are benign and beautiful (orchids and birds of paradise); some are large (elephants), others are small (hummingbirds and spider mites, never mind the vast numbers of organisms that we can only see under the microscope).

We don't know if Haldane *really* made his remark about beetles, and some may see the humor as one more way that science has often tried to belittle theology. But the comment highlights our ignorance of the natural world and the need to better understand organisms and their interactions with their complicated and often fluctuating environments, and how (or perhaps if) this relates to the ways life diversified across the planet over vast periods of time. There is a serious point: we do need some method of giving order to the vast number of biological sorts and a framework for understanding how they relate to each other.

Such a framework, incomplete as it is, is the subject matter of ecology. Formally, ecology is the study or knowledge (-*ology*) of places (*ecos*) and their inhabitants; it is about the "home life" of living organisms. A standard textbook definition is that "ecology is the scientific study of the interactions that determine the distribution and abundance of organisms." This statement seems simple and straightforward, but it is open-ended; ecology has no general rules like Newtonian physics or the Periodic Table in chemistry. Anyone who has watched nature programs on TV or visited natural habitats, or even gone to a large zoo, knows how enormous the subject matter of ecology can be—it forms a colossal complexity, set out in Figure 1.1.

The twenty most important concepts in ecology identified by professional ecologists are:[1]

1. The ecosystem
2. Succession
3. Energy flow
4. Conservation of resources
5. Competition
6. Niche
7. Materials recycling
8. The community
9. Life-history strategies
10. Ecosystem fragility
11. Food webs
12. Ecological adaptations
13. Environmental heterogeneity
14. Species diversity
15. Density-dependent regulation

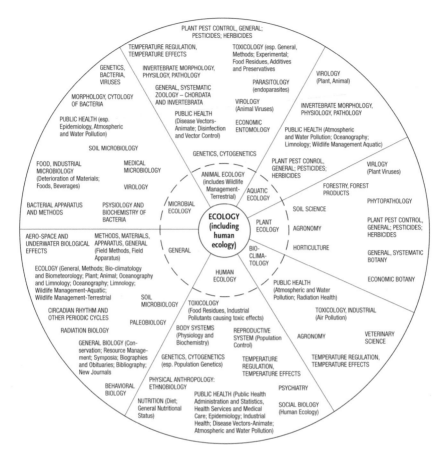

FIGURE 1.1. The centrality of ecology to other disciplines (from *Biological Abstracts Index*, permission of Thomson-Reuter).

16. Limiting factors
17. Carrying capacity
18. Maximum sustainable yield
19. Population cycles
20. Predator-prey interactions

In this brief survey of ecology, we can do no more than to sample just a few of these topics, choosing our illustrations carefully but hopefully avoiding over-simplification. More information and detailed analysis of these twenty concepts can be found in the many excellent textbooks available (see Further Reading at the end of this volume). All we can do here is to dive below them, following the ecologist Larry Slobodkin when he pointed out that "from the viewpoint of a cat, mice are machines for converting plants into food," referring to basic mechanisms—the interaction of animals with each other, their food, and ultimately their survival. What we are concerned with in this book is this core of ecology: describing where organisms occur, how many occur there, and most difficultly, why.

At this point, we need to recognize that "ecology" is not the same as "environment." The environment of an organism consists of all those factors outside the organism that influence it—both physical and chemical (*abiotic*) and other organisms (*biotic*). The "interactions" in the definition of ecology include, of course, the interactions with these factors. Environment is therefore inseparable from but wider than ecology. Nevertheless, while ecology can be regarded as no more than a specialized part of biology, in fact it overflows into an examination of questions about order and purpose in the world, questions that have repeatedly been asked through the centuries: Herodotus on predators, Basil on forest succession, problems about the size and inhabitants of Noah's ark, Benjamin Franklin speculating on the control of insect pests by birds, modern concerns about the effect of extinctions, and so on.

As scientific knowledge became increasingly ordered in the eighteenth and nineteenth centuries, it was perhaps inevitable for ecology to emerge as a distinct field of study. The word itself (ecology) was invented by a German biologist (Ernst Haeckel) as recently as 1866. His intention was to distinguish ecology as a discipline so that it could develop independently from other parts of biology (particularly the traditional studies of form and function—otherwise known as morphology or anatomy, and physiology), but the practice of ecology is really much older. Indeed it has been claimed as the oldest profession, since God gave Adam the task of naming the animals in paradise before he even created woman and the consequent possibility of sexual rambling.

However, the first sketch of ecology in the early days of science proper can be attributed to Carl Linnaeus (1707–1778) in a 1749 essay, *Oeconomia Naturae*, in which he used reproduction, cooperation, and mortality to describe the key elements in the "economy" of an organism. When combined with the awe and respect of the natural world in the writings of people like John Ray (1627–1705; *The Wisdom of God Manifested in the Works of Creation*, 1691), William Derham (1657–1735; *Physico-Theology*, 1713), and even John Wesley (1703–1791; *Survey of the Wisdom of God in the Creation*, 1763), we have the beginnings of a tradition of scientific natural history that developed in Britain into the pursuit of ecology, a tradition somewhat different from the more physiological approach that emerged in Germany and North America, although the end results are inevitably the same.

In its modern guise, ecology surfaced at the same time as genetics at the beginning of the twentieth century. Its first professional body was the British Ecological Society, founded in 1913 as the successor of a "Committee for the Survey and Study of Vegetation," a group of botanists who came together to use their passion for collecting wildflowers and swapping rarities as a means for determining the distribution and characteristics of different species and hence their limits and preferences. The Ecological Society of America was founded two years later.

Several factors contributed to forge ecology. Until the end of the seventeenth century, there was general agreement that the world was a mere five to ten thousand years old (Table 1.1). Doubts about this dating began to surface from cosmological speculations around the beginning of the eighteenth century.

In the 1790s the French mathematician, Pierre-Simon Laplace[2] (1749–1827) made explicit earlier suggestions that the planets were the result of the condensation of incandescent gas rotating around a central sun. The Comte de Buffon (1707–1788), director of the Royal Botanic Garden in Paris, took on the implications of this and calculated how long it would take a sphere to cool from white heat

TABLE 1.1 Proposed Dates of Creation

A. TRADITIONAL UNDERSTANDING	
Mayas (6th century BC)	3114 BC
Orthodox Judaism	3925 BC
Eusebius (early 4th century AD)	5799 BC
Bede (8th century AD)	3952 BC
Archbishop Ussher (1654)	4004 BC
Buffon (1779)	74,832 years old
Lord Kelvin (1898)	20–40 million years old
B. SCIENTIFIC UNDERSTANDING	Years before present
Origin of the universe	13,700 million
Origin of the solar system (= origin of Earth)	4,566 ± 2 million
Oldest known minerals on Earth (zircons)	4,408 ± 8 million
Oldest known rock on Earth	4,031 ± 3 million
Earliest evidence of life on Earth (carbon-13 depleted graphite)	3,850 million
Earliest microbial fossils on Earth	3,500 million
First cyanobacteria	2,000 million
First multicellular red algae	1,200 million
Oldest multicellular animal	575 million
First placental mammals	135 million
Earliest hominid (*Australopithecus*)	c. 5 million
Early modern *Homo sapiens*	c. 200,000
Neolithic agriculture	c. 12,000–10,000
First man on the Moon	AD 1969

to its present condition. He concluded that it would need 74,832 years. (He actually estimated that the Earth was at least half a million years old, but a previous book of his had been censored, so he kept quiet about this.) This stretching of Earth history was soon reinforced by the inferences of the early geologists William Smith (1769–1839), James Hutton (1726–1797), Charles Lyell (1797–1875), and others, who studied landscapes and the weathering of rocks. Their work led to the general acceptance that the Earth was hundreds of thousands of years old, although accurate estimates only became possible with the discovery and application of radioactivity to dating in the twentieth century. The lengthening of the Earth's age also meant that it might be different from its original state. Time allows change.

At the time, geological change was not linked to biological change ("transformism" in the language of the eighteenth century). The main reason was that there always seemed to be the possibility of new life appearing at any time through spontaneous generation; an apparently new form might actually be a newly arisen organism rather than a changed one. Speculating about biological change was not worthwhile. Aristotle taught that aphids came from dew, fleas from decaying matter, mice from moldy hay, and crocodiles from rotting logs. The wrongness of such ideas and the need to think about transformism only became urgent following the experiments in 1861 of Louis Pasteur (1822–1895), which showed that the apparent spontaneous occurrence of new forms was nothing more than contamination.

THE ORIGINS OF LIFE: "A WARM LITTLE POND"

Once the notion of spontaneous generation was ruled out, questions began to be raised about life: How did it arise? What is it anyway? Is its essence a self-replicating molecule, or is the step from self-replication to more complex life the real problem? In 1907 the Swedish chemist Arrhenius (1859–1927) proposed that life came from outer space as microorganisms drifted between planets, a notion called "panspermia." Remnants of this idea still persist, particularly with the finding that some fairly complex molecules are sometimes found in debris from space. However, most effort has been put into finding ways in which life may have originated on Earth. Charles Darwin suggested one possibility in a letter (1871) to his friend Joseph Hooker. He proposed that the original life may have begun in a

> warm little pond, with all sorts of ammonia and phosphoric salts, lights, heat, electricity, etc. present, so that a protein compound was chemically formed ready to undergo still more complex changes. . . . At the present day such matter would be instantly devoured or absorbed, which would not have been the case before living creatures were formed.

Aleksandr Oparin (writing in Russian) and, independently, J. B. S. Haldane took up this idea in 1924 and 1929, respectively. They suggested that the early seas might have functioned as Darwin's warm soup, containing relatively simple inorganic compounds (ammonia, methane, hydrogen sulfide, carbon monoxide, phosphate radicals) that did not immediately break down as they would nowadays because the atmosphere was (they assumed) oxygen-free. In due course, these simple chemicals might have combined into more complex molecules in various, random ways under the influence of energy from sunlight or lightning. Stanley Miller and Harold Urey tested this hypothesis in 1952 at the University of Chicago. They passed an electric current through a sealed vessel containing methane, ammonia, hydrogen, and water. After a week they found eleven of the twenty commonly occurring amino acids (the building blocks of proteins) in their solution. In 1961 Joan Oró, a Spaniard working with NASA, found in a similar experiment that bases which are part of nucleic acids appear where hydrogen cyanide is in the solution.

Much debate surrounds these experiments. We cannot be certain of the exact nature of the early atmosphere. More important, we do not know how the simple organic molecules turned into self-replicating systems. Some scientists have claimed that juxtaposition in bubbles or on clay could have catalyzed their assembly. Such systems may have been helped by the sort of extreme conditions found in volcanoes or deep sea vents or by radioactivity. We have to acknowledge that all we can do is speculate. But at least most people regard the speculations as plausible.

One person who disagreed that life originated on Earth was the astronomer Fred Hoyle. He was a proponent of panspermia, arguing that the idea that life originated by the shuffling of molecules in a primeval soup was "as ridiculous and improbable as the proposition that a tornado blowing through a junk yard may assemble a Boeing 747." He estimated that the probability of life beginning from a random assembling of molecules to be one in ten to the power of forty thousand—the chance that two thousand enzyme molecules would be formed simultaneously from their twenty component amino acids on a single specified occasion. But this calculation was not correct: the relevant chance is of some far simpler system being formed at any place on the Earth at any time within 100 million years or so. We cannot calculate this probability, because we know neither the nature of the hypothetical system nor the composition of the primeval soup in which it arose. The origin of life was obviously a very rare event, but we have no intrinsic reason to think it as extra-ordinary as Hoyle calculated.

We are on rather surer ground when we talk about the origin of simple organisms.

HISTORY OF LIFE ON EARTH

A chain of being, or *scala naturae*, was assumed from early times in European culture, its ordering determined by decreasing spiritual status: from God down through other heavenly beings, to humans (as uniquely invested with souls among earthly creations), thence through the animal kingdom, to plants, and finally inorganic rocks. For centuries, this chain of being was assumed to be a permanent and fixed arrangement laid down at creation, but in the late eighteenth century evolutionary ideas began to circulate. The chain of being was changed into a basis for speculating about relationships, stimulated to no small extent by Charles Darwin's grandfather, Erasmus (1731–1802). Perhaps the best-known early evolutionist was a French zoologist, Jean Baptiste Lamarck (1744–1829), who proposed in his *Philosophie zoologique* of 1809 that simple life arises continuously from inanimate matter and progressively evolves toward "higher" forms. Lamarck was motivated in part by theological concerns to explain away the possibility of extinction in a "good" creation. He believed in an inevitable upward progression with a special place for humankind.

This naïve idea seemed at first to be supported by paleontology, when the fossil record first began to become clear in the early decades of the nineteenth century. The early paleontologists believed that life began with a big bang in the Cambrian era, which lasted from around 542 million years ago to 488 million years ago. The earliest Cambrian rocks contain fossils of primitive arthropods (trilobites) and an explosion of early forms of virtually all the major groups currently alive. However, in the last few decades, "stromatolite fossils" have been described in much older rocks at Warrawoona in Australia, aged 3,450 million years. Fossils (*Grypania*) of more complex eukaryotic cells, the base from which all animals, plants, and fungi are built, have been found in rocks from 1,400 million years ago in China and Montana. Other rocks dating from 565 to 543 million years ago contain fossils of the Ediacaran biota (named after the Ediacara Hills of south Australia), organisms so large that they must have been multicelled, but very unlike any modern organisms. The Cambrian explosion seems to mark not so much the beginning of life as the time when hard skeletons appeared, which left recognizable fossils.

From the Cambrian times onward, we have a record of the history of life showing ever increasing diversification and complexity (Table 1.2). The Paleozoic era from the start of the Cambrian to the end of the Permian 251 million years ago was a time when fishes dominated, but also when amphibians invaded the land, when reptiles appeared and spread, and when massive forests covered large areas, forming in due course coal deposits. The Paleozoic ended with a major extinction event and the loss of much of the marine fauna. The Mesozoic era came next, the age of the dinosaurs and the appearance of flowering plants. The Mesozoic also ended with a major bout of extinctions, commonly attributed to a meteorite impact 65.5 million years ago. This event is commonly regarded as occurring at the K-T boundary (from the German words for "Cretaceous" and "Tertiary"); it marks the disappearance of the dinosaurs and the beginning of the modern era, including the spread of mammals and finally the appearance of human beings.

The fossil record (or "tree of life," as it is sometimes called) was recognized by the early nineteenth century, although the geologists of that time had no means of dating the rocks or the fossils contained in them. However, the general acceptance of the fossil record gave strength to the notion that the world was much older than traditional assumptions, clearing the way for the more sophisticated evolutionists who followed Lamarck. Darwin believed the Earth was about 300 million years old. In 1904, less than ten years after the discovery of radioactivity, Ernest (Lord) Rutherford suggested that radioactive decay could be a method of dating the age of rocks. Techniques developed over the next half century; by 1953 the age of the Earth was calculated to be 4.55 billion years, and the age of life on Earth to be 3.5 billion years. More than forty radioactive breakdown series can now be used independently of each other to determine the age of rocks and fossils. No major discordances exist between dates based on different methods.

TABLE 1.2 Geological Column

	Millions of years ago (mya)*		
Cenozoic Era (Age of Recent Life)	Quaternary Period	The geologic eras were originally named Primary, Secondary, Tertiary, and Quaternary, but only Tertiary and Quaternary have been retained, and used as period designations.	Ice ages and modern times
	2.6		
	Tertiary Period		Age of mammals
	65		
Mezozoic Era (Age of Dinosaurs)	Cretaceous Period	Derived from Latin word for chalk (*creta*) and first applied to the extensive deposits that form white cliffs along the English Channel.	Last dinosaurs, first primates, first flowering plants
	146		
	Jurassic Period	Named for the Jura Mountains, located between France and Switzerland, where rocks of this age were first studied.	Dinosaurs dominant, first birds
	200		
	Triassic Period	Taken from the word *trias* in recognition of the threefold character of these rocks in Europe.	Appearance of dinosaurs Extinction of many marine forms

*The dates are the time before the present in millions of years at the beginning of each period.

	251		
	Permian Period	Named after the Russian province of Permia, where these rocks were first studied.	Widesprad deserts
	299		
Paleozoic Era (Age of Ancient Life)	Carboniferous Period	Extensive coal-bearing deposits.	First reptiles, seed ferns
	359		
	Devonian Period	Named after Devonshire, England, where these rocks were first studied.	First amphibinas. jawed fish
	416		
	Silurian Period	Named after Celtic tribes, the Silures and the Ordovices, that lived in Wales during the Roman Conquest.	First vascular land plants
	444		
	Ordovician Period		
	488		
	Cambrian Period	Taken from the Roman name for Wales (Cambria) where rocks containing the earliest evidence of complex forms of life were first studied.	Most modern phyla appear.
	542		
Pre-Cambrian	Proterozoic Archean	The time between the birth of the planet and the appearance of complex forms of life. More than 80 percent of the Earth's estimated 4.5 billion years falls within this era.	

CHANGING CLIMATES

The Earth has changed in many ways since its formation. At one time it seemed inconceivable that the continents were not in the same position as when they were created. Then in the 1880s an Austrian geologist, Eduard Suess, speculated about massive rising and falling of sea levels. He was impressed by the discovery that the seed fern *Glossopteris* occurred widely in South America, South Africa, Australia, and Antarctica. He argued that those continents might once have been part of a single landmass that he called Gondwanaland (after rocks in central northern India from which the Gondwana sedimentary sequences are named—the "forest of Gond"). In 1912 a German, Alfred Wegener, built on this, suggesting that the continents might actually have moved. He argued that this could explain the uneven distribution of living things on our planet— why Australia has no true mammals, for example. Wegener proposed that there was once only one single supercontinent, which he called Pangaea ("all land"). The implication was that Pangaea split to yield Gondwanaland and another large mass called Laurasia (the name being a compound of the St. Lawrence River and Asia).

Such notions were widely ridiculed; they seemed intrinsically absurd. This scoffing was muted when during the Cold War, submarines began to explore the ocean floors and found new material spreading from mid-ocean ridges, which had been discovered by cable-laying ships. Some credence began to be given to the idea that continents might indeed move. All this led by the 1960s to the acceptance of the existence of tectonic plates and the recognition that the Earth's surface (or lithosphere) consists of eight major and around sixty minor plates that may diverge or collide, leading to earthquakes, volcanic activity, and mountain building. Suess's conclusion that the Alps were once on the sea floor has been proved right, albeit through a mechanism of which he had no idea.

The current belief is that an original supercontinent (Rodinia—

from the Russian word for "motherland") emerged about a billion years ago, then split into eight continents around 750 million years ago. These eight reassembled into Pangaea, which in turn divided in the Jurassic era 180–220 million years ago into Laurasia (which became North America and Eurasia) and Gondwana (which included Antarctica, South America, Africa, Madagascar, Australia–New Guinea, and New Zealand, as well as Arabia and the Indian subcontinent). When Gondwana broke up, East Gondwana (comprising Antarctica, Madagascar, India, and Australia) began to separate from Africa, while South America began to drift away, opening the South Atlantic Ocean from about 130 million years ago (see Figures 1.2a and 1.2b).

As these movements were taking places, ever-evolving plants and animals began to populate the land.

CLIMATE HISTORY

Over the vast spans of prehistoric time, the Earth has alternated between extreme cold and extreme heat (Figure 1.3). In the first three-quarters of the Earth's history, only a single major glaciation has been found, but since around 950 million years ago, the Earth's climate has varied every 140 million years or so between large (global) or small (just polar-cap wide) glaciations and extensive tropical climates. This variation may be related to Earth's motion into and out of galactic spiral arms and significant changes in solar winds.

Climates from the distant past can be reconstructed from the isotope proportions in rocks. One requirement for the development of large-scale ice sheets seems to be the arrangement of continental landmasses at or near the poles. In other words, the constant rearrangement of continents by plate tectonics can influence long-term climate change. However, the presence or absence of landmasses at the poles is not in itself sufficient to guarantee glaciations or exclude polar ice caps. Evidence exists of past warm periods in

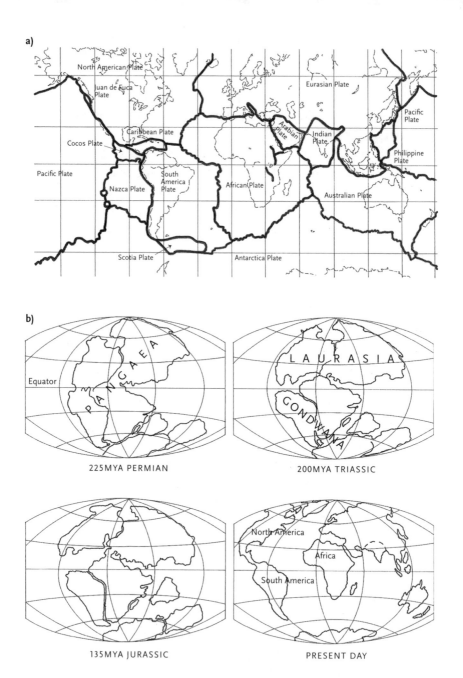

FIGURE 1.2. Continental drifting. a) The surface of the Earth (lithosphere) is composed of tectonic plates, which move over the more liquid subsurface layer (asthenosphere). **b)** The Earth's continents have drifted over geological time, joining in supercontinents and then splitting before arriving at their present position.

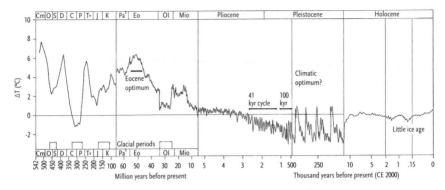

FIGURE 1.3. Temperatures over the last half million years (note that the horizontal scale is logarithmic).

Earth's climate when polar landmasses similar to Antarctica were home to deciduous forests rather than ice sheets.

In the late Precambrian the continents were joined together in the Rodinia supercontinent with major glaciations extending over much of the earth. Massive deposits of rocks formed from glacial till are found from that time, giving rise to the idea of a Snowball Earth. General warming followed the cold of the Precambrian, and by the beginning of the Cambrian average global temperatures were about 22°C (72°F). "Global mean temperature" is a difficult concept to absorb. We commonly experience differences in temperature between day and night of the order of 10°C (18°F) or more. Perhaps surprisingly, the difference in global mean temperatures between a fully glaciated Earth and a completely ice-free Earth seems to be a similar 10°C , though larger changes occur at high latitudes and smaller ones at low latitudes. As we will see, very small changes in temperature can have a widespread impact on the Earth's surface if they are sustained.

During the Jurassic and Cretaceous, the Rodinia supercontinent broke up and the Earth was relatively warm. Superimposed on the long-term evolution between hot and cold climates were probably many short-term fluctuations in climate, similar to and sometimes more severe than the varying glacial and interglacial states of recent

Ice Ages. Some of the most severe fluctuations, such as a period of temperature rise at the Paleocene-Eocene boundary when global average temperatures rose by 6°C (11°F) over twenty thousand years, may be related to rapid climate changes due to sudden emptying of natural methane ("clathrate") reservoirs in sediments below the oceans. Other severe climate changes are recorded at the Cretaceous-Tertiary, Permian-Triassic, and Ordovician-Silurian boundaries, and are associated with major extinction events.

The present (Quaternary) era stretches to include the current climate. Nonetheless a succession of ice ages has taken place over the last 2.2 million years or so. These show a strong periodicity of about every one hundred thousand years, with strikingly asymmetric temperature curves (Figure 1.4). This asymmetry is believed to result from complex feedback mechanisms.

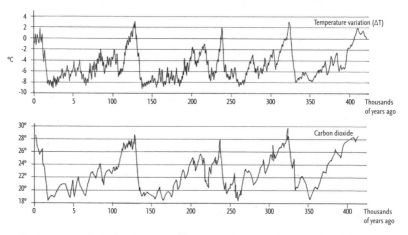

FIGURE 1.4. Variation in temperatures and atmospheric carbon dioxide during the Quaternary Period, based on gas bubbles trapped in an ice core drilled at the Vostok Station in the Antarctic. Note that the temperature and carbon dioxide curves are effectively identical.

In general, ice ages deepen by progressive steps, but the recovery to interglacial conditions apparently occurs in one big step. These cycles are thought to be driven by so-called Milankovitch cycles, named after the Serbian civil engineer and mathematician Milutin Milanković, who argued that variations in eccentricity and axial tilt ("precession") of the Earth's orbit influence climatic patterns on Earth. In effect, the Earth wobbles on its axis, causing its surface to move closer to, or farther from, the Sun, with consequent changes in climate. The Earth's axis completes one full cycle of precession approximately every twenty-six thousand years. At the same time, its elliptical orbit rotates more slowly, leading to a twenty-three-thousand-year cycle between the seasons and the orbit. In addition, the angle between Earth's rotational axis and the plane of its orbit moves from 22.1 degrees to 24.5 degrees and back again on a forty-one-thousand-year cycle.

The position of the continental plates on the surface of the Earth also affects the timing of ice ages throughout geologic history. When landmasses are concentrated near the polar regions, there is an increased chance for snow and ice to accumulate. Small changes in solar energy can tip the balance between summers in which the winter snow mass completely melts and summers in which the winter snow persists until the following winter.

LAST ICE AGE AND AFTER

Our present geological era (the Quaternary) is divided into Pleistocene and Holocene epochs. Charles Lyell (1797–1875) coined the name "Pleistocene" in 1839 to describe rock strata containing the fossils of mollusks where 70 percent or more of the fossils belong to species that are still alive. The Pleistocene covers the period we rather loosely call the Ice Age, although it was actually a time when a series of cold periods was interspersed with warm interstadials. The events of the Pleistocene are largely responsible for the appearance of the present landscapes of the temperate areas of North America and northern Eurasia. Erosion by glaciers molded the highlands, while a mantle of deposits from them covers lower ground.

The glaciers of the last Ice Age covered most northern latitudes. They began to retreat about 18,000 years ago. Another flurry of cold occurred briefly between 12,900 and 11,550 years ago, but this was followed by a general warming, which ushered in our current Holocene epoch. *Homo erectus*, the immediate predecessor of *Homo sapiens*, appeared on Earth more than a million years ago. Over the past three-quarters of a million years, our ancestors have lived through eight, perhaps nine, ice ages. During this time our present flora and fauna began to acquire its familiar characteristics. As the ice melted, sea levels rose. Britain became separated from continental Europe.

As the last Ice Age glaciers retreated, temperatures rose fairly rapidly. Between 9,000 and 5,000 years ago, summer temperatures were about 2–3°C (4–6°F) higher than at present. Pollen deposits preserved in bogs show that woodland was widespread in areas freed of ice; birch, hazel, and elm, preceding colonization by oak and ash. The climate in England is evidenced by archaeological studies of settlements and farming in the Early Bronze Age at heights now above cultivation, such as Dartmoor, Exmoor, the Lake District, and the Pennines. Settlements and field boundaries have been found at high altitude in areas that are now wild and largely uninhabitable. During this period cave art and other signs of settlement in prehistoric Central North Africa show that some parts of the present Saharan Desert may have been inhabited. Temperatures dipped again in the Late Bronze Age around 5,000 years ago, although not as much as in the ice ages proper. Elm declined rapidly throughout the Northern Hemisphere over a period of only 50 to 100 years (see Figure 1.5).

Temperatures rose again in the first millennium after Christ, with another warm period in the north between about AD 800 and 1300. Greenland was colonized by Scandinavian voyagers, who established a colony along the lines of their traditional culture in Norway. But several centuries later, temperatures fell again. Sea ice spread southward, and by 1203 Norse ships had difficulties sailing from Iceland. Early frosts and crop failures in Poland and on the Russian plains in 1215 caused famine, leading people to sell their children and eat pine bark. This contrasted with a period of unusu-

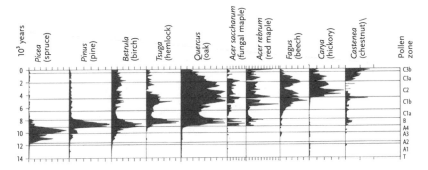

FIGURE 1.5. Changes in tree species in Connecticut from the end of the last glacial period to the present, as shown by concentrations of pollen collected from sediments in Rogers Lake, Lyme (based on Margaret B. Davis, "Pollen Accumulation Rates at Rogers Lake, Connecticut, during Late- and Postglacial Time," *Review of Palaeobotany and Palynology* 2 [1967]: 219–30).

ally warm and mainly dry summers between 1284 and 1311; there were vineyards in England. Then all changed again. Continuous downpours in 1315 meant the harvest was a disaster. The Greenland Vikings could not sustain their lifestyle. Their inflexibility in a changing climate proved their death knell.[3] Their settlements disappeared, although the native Inuit were able to survive in the same area.

The events of the early fourteenth century marked the beginning of a Little Ice Age that lasted until the mid-nineteenth century. Winter fairs were held on the frozen Thames. Mountain glaciers advanced far beyond their modern range in the Alps, Iceland and Scandinavia, Alaska, China, the southern Andes, and New Zealand. Snow lines were 350 meters (1,150 feet) below modern levels and close to their position in the last Ice Age proper; glaciers extended more than 100 meters (330 feet) lower than at present.

The effects of these changes in climate varied from place to place. From the 1630s the Ming Empire of China faced widespread drought. In the 1640s drought in the fertile Yangtze River valley was followed by catastrophic floods, epidemics, and famine. Millions of people died from hunger and the wars that followed the fall of the dynasty in 1644. Eastern North America had its coldest weather in the nineteenth century, but the western United States was warmer than in the mid-twentieth century. Around 1850 the climate began warming generally, and the Little Ice Age ended. Some believe that Earth's climate is still recovering from the Little Ice Age and that human activity is not the decisive factor in present temperature trends, but regional differences complicate interpretations and this skepticism is not widely accepted.

HUMAN HISTORY

Charles Darwin's friend and geological mentor, Charles Lyell, long resisted evolutionary ideas, particularly as they related to human beings. He regarded humans as too distinctive to have evolved by natural selection. However, Lyell was reluctantly persuaded by Darwin's arguments in the *Origin of Species* that we share a common ancestry with the great apes. He even went as far as to suggest in his book *The Antiquity of Man* (1863) that the best place to look for human fossils would be in Africa where gorillas and chimpanzees live today. He has been proved spectacularly right. Diligent searching has led to the discovery of a considerable number of fossil hominids in Africa, to the extent that we can claim to have the best fossil history of any species.[4]

All the remains of early, prehuman hominids have been found in Africa. Outside Africa, there are fossils of *Homo* species in many parts of southern Europe and Asia dating from 1.8 million years ago and later. Our own species, *Homo sapiens*, seems to have originated in Africa and spread from there at a rather late date, probably initially about two hundred thousand years ago. We are all descended from the same stock; we are all members of the same species. Non-Africans are less genetically variable than Africans, and this is one of the main pieces of evidence for the idea that a group of prehumans moved north from Africa into the Middle East and then much of the Old World. This migration out of Africa took place sixty-five thousand to forty thousand years ago, groups establishing settlements wherever they found food and shelter.

Our early ancestors shaped and used stone tools. They are properly called "Stone Age" people. Then around eleven to nine thousand years ago signs of domesticated crops and animals begin to appear in archaeological sites. in North Africa, China, and other parts of south Asia, but most significantly in the Middle East. Domestication—particularly of plants—implies that the domesticators were living a settled existence. There is no point in sowing a crop and then moving somewhere else. Domestication means the beginning of farming, a move from hunting and scavenging to a relatively settled way of life—in archaeological terms, a change from Paleolithic to Neolithic.

The pieces of the jigsaw that makes up the modern world were in place.

CHAPTER 2
A Green Machine

In order to make an apple pie from scratch,
you must first create the universe. —Carl Sagan

THE NATURAL WORLD is a vast green machine, a colossal engine
fueled by energy from the Sun. It is a remarkable system: plant life
acquires energy for growth and reproduction , and then becomes
food for animals. The whole of the living world can be regarded
as an enormous array of intermeshing and interacting parts, some
of them simple, most very complicated, and all regularly sur-
prising. Ecology is about dissecting this mechanism, using every
available technique to study animals and plants in their wider
environment.

The data about the natural world are so huge that they can be
overwhelming. Like any other science, ecology has to decipher this
complexity as much as possible, while searching for a few simple
principles and rules. As this chapter shows, Charles Darwin gave
us one of the most important ecological principles in his recognition of natural selection as a mechanism for producing adaptation.
Other generalities have followed. In recent years, new technology such as automated loggers (which count off natural events)
and satellites that download remotely sensed data contribute to an
impressive tool kit to classify life processes and study an enormous
range of events—from the way predation works to the formation
of "niches" and how food webs operate in nature.

It is often difficult to separate the relevant from the trivial, background variation from real trends. The complete environment of any animal or plant tends to be bewilderingly complex. It involves the structure of the individual itself, as well as the physical and chemical components of its surroundings (temperature, humidity, acidity, salinity, etc.). It includes the influence of all the other organisms in the vicinity: individuals of the same species that may be competing for resources, and individuals of other species—which may afford a hazard or an opportunity (predators, parasites, or pathogens; food or hosts). And on top of all this, it has to incorporate the history and vicissitudes of former occupants of the area.

In practice, this complexity leads to ecologists concentrating on particular parts of the puzzle: the study of individuals within a particular species (*autecology*) or groups of species (*synecology*). Synecology in turn can be approached at the population, community, or ecosystem level; it involves bringing together both form (anatomy or morphology) and function (physiology—which may be at the biochemical, cellular, or some higher level). Some studies are top down, as it were, asking about the influence of environmental factors on the success or failures of particular groups or species. Species vary in their tolerance of temperature fluctuations (such as frost sensitivity) or light or pollutants, and so on. This leads to questions about geography: Why is a particular species or group of species where it is (or isn't)? How do species or groups spread? What are the requirements for their survival? Genetics has becoming increasingly important in ecology as new techniques show that groups (or populations) of animals and plants are surprisingly variable and different in their responses to their environment. Ecology shades into geography and evolution; its boundaries are wide (see Figures 1.1 and 2.2).

In the early days of ecology as a recognized study, most of those involved took it for granted that communities of animals and plants existed as natural, repeated, internally organized units, with a considerable degree of integration. Such biotic communities were treated as superorganisms, inheriting the common medieval analogy between microcosm and macrocosm that became assimilated into the idealism of Immanuel Kant and Johann von Goethe in Germany. In a similar way, the pioneering American botanist Frederic Clements (1874–1945) argued strongly that this integration implied an inevitable development toward a predestined end point, or "climax community." His most influential book (*Plant Succession*) was published in 1916.

In the 1930s similar ideas came from zoologists, particularly from a pioneering group of ecologists in Chicago who were puzzled by the division of labor in colonial hymenoptera (bees, ants, termites) since all the individuals were born to a single queen and were therefore genetically identical. Like Clements with his plant communities, perhaps such colonies were really "superorganisms," with the proportion of different castes regulated by some sort of homeostatic feedback.

In the botanical context, Clements's ideas were taken up by John Phillips, an Edinburgh-born botanist. Phillips argued that living communities were more than the sum of their parts and could be seen to have a destiny peculiar to themselves. Such emerging mysticism was anathema to the leading British ecologist of the time, Arthur Tansley (1871–1955). He sought to defuse it by coining a new word, "ecosystem," which included "not only the organism-complex, but also the whole complex of physical factors forming the habitat factors in the widest sense. It is the systems so formed which, from the point of view of the ecologist, are the basic units of nature on the face of the earth." His hope was that purging the superorganism concept from the scientific vocabulary would open plant (and, indeed, animal) communities to scientific investigation.

Sadly, ecosystems, which Tansley intended to represent as a simple descriptive generality, have spawned their own subdiscipline with ascribed properties of resilience, persistence, resistance, and variability. In this wrongful extension of Tansley's idea, ecosystems have become synonymous with "local biological" situations. Ecosystems are commonly said to have "health" and "needs" and to suffer damage (designations properly attributed to organisms).

This abuse of the concept is an unjustified extrapolation. It is preferable to restrict the description of an ecosystem to *a self-organizing system in which random disturbance and colonization events create a heterogeneous landscape of diverse species, which then become knitted together through nutrient fluxes and other forms of interaction, some simply having to do with chance and geography.* In other words, we should confine "ecosystem" to the neutral idea that Tansley intended, and bring "ecology" back to a concern with the distribution and abundance of organisms—that is, the study of living communities and their interactions—while recognizing a wider and less precise area of "environmental studies," which incorporate physical influences, including geological and climatic factors.

Another implication of the superorganism idea rumbles on today. If a biological community behaves as an entity, could the Darwinian idea of natural selection be applied to groups, and not just to individual organisms? As we shall see, this would be able to explain the existence of self-sacrificial behavior, which was (and remains) a nagging problem. Even more radically, it could challenge the assumption that nature was irredeemably "red in tooth and claw." A possible answer came in the 1920s from J. B. S. Haldane (1892–1964). He proposed that it could be in their self-interest for nonbreeding individuals to defend the offspring of their closest relatives (in other words, altruism), even if they did not pass on their own genes. Altruistic behavior could be compatible with the Darwinian mechanism through this particular version of group selection. All this remained little more than speculation until the pioneering work of Bill Hamilton (1936–2000) nearly half a century later. Altruism as little more than genetic self-interest was finally popularized as "sociobiology" by Edward O. Wilson (1929–) (see p. 145).

ADAPTATION AND VARIETY

Probably the most common way of describing the interaction between an organism and its environment is to say, "Organism X is adapted to . . ." Fish are adapted to living in water, cacti to living in deserts, and so on. In past centuries this was assumed to be the result of a creator's design, making a world in which everything had its own place and function. The classic statement of this understanding was set out in *Natural Theology,* published in 1802 by an English clergyman, Archdeacon William Paley (1743–1805). In a famous analogy, Paley reasoned that a watch would be expected to have had a watchmaker, whereas there would be no grounds for assuming that a stone was designed. It followed that our wonderfully complex natural world must have been designed; God was a great Watchmaker. Paley's writings were extremely influential in the early nineteenth century. Charles Darwin was required to read them while a student at Cambridge University. He wrote in his *Autobiography,* "The careful study of these works was the only part of the Academical Course which was of the least use to me in the education of my mind." He re-read Paley in the early 1840s when he first wrote down his ideas about evolution.

However, Paley's thesis did not long stand critical scrutiny, particularly in the light of the stretching of Earth's history, which was taking place even as Paley was writing (see p. 9). A creator could presumably design an organism perfectly adapted to a particular environment, but this perfection would disappear if the environment was not constant. Adaptation to changes in climate, to the physical structure of the Earth's surface, or to predators and competitors is possible only if organisms change. A long Earth history is an Achilles heel for traditional natural theology.

The way in which adaptation could take place (whether or not the mechanism was divinely overseen) was Darwin's chief contribution to science, presented in the *Origin of Species,* published in 1859. He had read "for amusement" Thomas Malthus's (1766–1834)

Essay on the Principles of Population, which set out the specter of the human population outstripping its food supply, leading to the weak and improvident succumbing in the consequent struggle for resources. Darwin's genius was in linking the struggle for existence described by Malthus—an ecological concept (see p. 4)—to the fact of heritable variation. If only a small proportion of a population survives the struggle, the likelihood will be that it will include those with a trait giving its carriers some sort of advantage. Over the generations, the proportion of those individuals with the trait will increase at the expense of those lacking it. A genetic change would take place in the population, amounting to natural selection for the trait in question. In other words, natural selection leads to adaptation.

The easiest way to study natural selection in the natural world is to concentrate on a particular species, because there we can compare the survival or reproduction of different variants. One of the clearest examples of such a study was the spread in Britain of black forms of tree-trunk-sitting moths as pollution in the nineteenth century killed the lichens on their normal resting places, making the insects conspicuous to predatory birds. This spread of black (melanic) forms began more than a hundred years or so ago, but then reversed as smoke control legislation resulted in the conditions for a regrowth of the lichens (Figure 2.1).[1] Other well-studied cases of natural selection include the spread of antibiotic-resistant bacteria, pesticide-resistant insects, visual bird predation on the snail *Cepaea nemoralis,* chromosomal variants with altitude and season in the fruit-fly *Drosophila pseudoobscura,* and metal tolerance in grasses. The examples are legion.[2] The effect of such selection is to produce local races (ecotypes) adapted to a heterogeneous environment—one where conditions vary with either time or space.

On a larger scale, a number of patterns emerge in animals. These are sometimes described as "ecological rules." In fact, they are nothing more than the outworking of physical principles, and have many exceptions. Here are just three examples:

a)

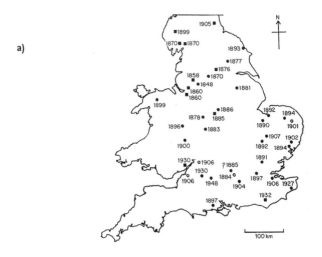

b)

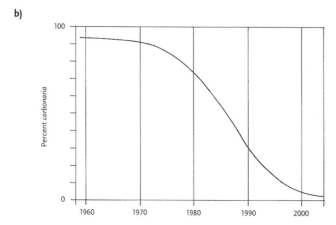

FIGURE 2.1. a) First records of the melanic form (*carbonaria*) of the peppered moth in Britain, spreading from its earliest finding in Manchester in 1848. Industrial pollution led to the disappearance of lichens from the resting sites of the moth; the environment of the moth thus changed to favor the melanic over the previous cryptic form. Circles show first recorded date; squares indicate dates when the form was already at an appreciable frequency (after R. C. Steward, "Industrial and Non-industrial Melanism in the Peppered Moth, *Biston betularia* [L.],"*Ecological Entomology* 2 [1977]: 231–43). b) Decline of *carbonaria* at Caldy, near Liverpool in northern England following the UK Clean Air Act of 1956 (based on data in Laurence M. Cook and J. R. G. Turner, "Decline of Melanism in Two British Moths: Spatial, Temporal, and Inter-specific Variation,"] *Heredity* 101 [2008]: 483–89).

▸ *Bergmann's Rule* states that body size of warm-blooded animals tends to be greater in colder regions. The amount of surface area of an individual decreases relative to mass, so a larger animal uses proportionately less energy keeping up its body temperature than a smaller one.

▸ *Allen's Rule* has the same basis: the extremities (ears, snout, legs) of warm-blooded animals are shorter in cooler parts of a species range than in warmer parts of that range.

▸ *Gloger's Rule*: Dark pigments are commoner in animals living in warm and humid habitats.

When it comes to adaptation, another strong and important rule is *convergence*, the occurrence of very similar structures or functions in different organisms living in similar habitats. Around the world, creatures with no link to each other have developed similar organs or body shapes (see Figure 2.2). These parallel forms must have originated in response to similar environmental pressures or opportunities; they are the results of adaptation to a similar end despite very different starting points. The best-known example of this is the occurrence on different continents of marsupial mammals (kangaroos and their relatives), which have an external pouch for the young (rather than the internal placenta of true mammals) paralleling true (or eutherian) mammals. Marsupial wolves, cats, flying squirrels, and anteaters have existed in Australia and South America, although most of these are now extinct in South America following the invasion of placental mammals from North America. The adaptations in the placental and the marsupial mammals must have developed from similar adaptive pressures.

On a smaller scale, the much-studied finches of the Galapagos Islands have diversified from a presumed generalist finch that managed to reach the islands (there is a putative relative on the South American mainland) into a range of different specialities (insect-eating, fruit-eating, woodpeckerlike, and so on) by exploiting opportunities only available to them because of the absence of species that would have filled these niches in a continental situation (see Figures 2.3a and 2.3b).

Another line of evidence for adaptation comes from the study of particular organs. All organisms benefit from being able to sense their environment, since only then can they react appropriately.

NICHE	PLACENTAL MAMMALS	AUSTRALIAN MARSUPIALS
Burrowers	Mole	Marsupial mole
Anteater		Numbat
Mouse		Marsupial mouse
Climber		Spotted cuscus
Glider	Flying squirrel	Flying phalanger
Cat	Ocelot	Dasyurus
Wolf	Wolf	Tasmanian wolf

FIGURE 2.2. Parallel evolution in placental mammals and marsupials. Very similar forms have developed independently in the two lines.

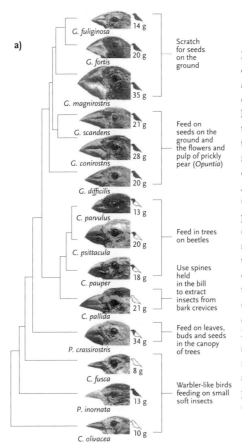

a)

G. fuliginosa — 14 g
G. fortis — 20 g
G. magnirostris — 35 g

Scratch for seeds on the ground

G. scandens — 21 g
G. conirostris — 28 g
G. difficilis — 20 g

Feed on seeds on the ground and the flowers and pulp of prickly pear (Opuntia)

C. parvulus — 13 g
C. psittacula — 20 g

Feed in trees on beetles

C. pauper — 18 g
C. pallida — 21 g

Use spines held in the bill to extract insects from bark crevices

P. crassirostris — 34 g

Feed on leaves, buds and seeds in the canopy of trees

C. fusca — 8 g
P. inornata — 13 g
C. olivacea — 10 g

Warbler-like birds feeding on small soft insects

FIGURE 2.3. a and b) The finch species of the Galapagos have become a classical example of adaptation to different habitats. The archipelago lies about 1,000 km (620 miles) off the coast of Ecuador and is home to thirteen species of finches found nowhere else in the world; their closest relative is on the South American mainland. The single Cocos Island to the north has no possibility for isolation on different islands and has only one finch species. The horizontal lines in the figure are measures of genetic distance derived by study of variation in DNA microsatellites (based on Kenneth Petren, B. R. Grant, and P. R. Grant, "A Phylogeny of Darwin's Finches Based on Microsatellite DNA Variation" *Proceedings of the Royal Society of London*, series B, 266 [1999]: 3321–29, by permission of the Royal Society of London).

b)

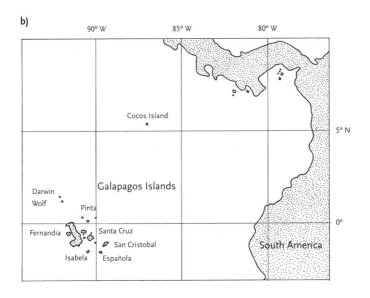

An important facility is the ability to detect light. Even very simple organisms have light-sensitive spots. More complex animals have either camera-type eyes (as in vertebrates—including humans) or compound eyes (typical of most insects) with a large number of simple elements. Squids and octopuses have camera eyes, but they have a completely different developmental origin and nerve supply to those of vertebrates. They have achieved a similar result but by a different route. Still more remarkable are the lenses that improve the capability of eyes. Lenses—all with the same function—have appeared in at least fifteen different ways in different organisms, co-opting different genes in their manufacture.

CLASSIFICATION OF FORMS

As soon as we start comparing different organisms, we need to make sure that we know what organisms we are discussing. We need to be able to identify and label different forms. So far we have assumed that a "species" does not need defining, but at this point it is necessary to digress and clarify what we mean by different forms. We are familiar with cardinal birds, common murres, redwoods, many sorts of oak trees, and giant pandas. There is truth but a degree of arrogance in saying that we can always distinguish species. In general most of us are able to recognize particular species, such as the difference between a tiger and a house cat. But what is a "species"? Spaniels and terriers are dogs, but biologists regard them as mere varieties of a domesticated form of the gray wolf, *Canis lupus*. What are the bounds of a species?

The most commonly accepted definition of a species is *a group of actual or potentially interbreeding populations*. Sometimes an animal or plant that looks like a "good" species may really be two or more different species which look so alike that they can only be distinguished by microscopical or biochemical techniques. Such *cryptic species* are nonetheless good species. For example, the malaria-transmitting mosquito originally called *Anopheles maculipennis* is actually a cluster of six different species that can only be separated by characters in their eggs. Nevertheless, they all have their own names.

Just as important is being able to recognize that a species is a biological unity even when it consists of a collection of races that may look different from each other. The story of the Orkney vole, a European relative of the deer mouse, illustrates this difficulty of recognition. When a "new" species of vole was discovered in the Orkney Islands, great excitement ensued—but then quickly disappeared when it was realized that the Orkney vole was merely a large variant of a species widespread in continental Europe.

This sort of situation causes difficulties for museum taxonomists because they have no way of knowing if the dead specimens which they handle are members of the same species. There are even examples where different sexes or different life stages have been classified as different species. That is a practical problem: the key is accepting that a species is really a collection of local forms that occupy the same niche in the natural world. All members of a species may look alike (*monotypic*), but there are many examples of *polytypic* species. The human species (*Homo sapiens*) is an example of a polytypic species. There are black, brown, yellow, and white skinned races, but we all belong to the same species, and members of all the races can successfully breed with each other.

All this makes language and accurate naming essential to avoid confusion. Such naming in biology forms the discipline of taxonomy. It involves a hierarchy of terms, moving from subspecific categories (form, race, variety), through the basic unit (the species), to the broadest groupings at the roots and trunk of the tree (or bush) of life. Higher categories (ones potentially containing different species, although some higher groups only contain one species) are genus, family, order, class, phylum, and finally kingdom.

Every species has a two-word name (or *binomial*), indicating its genus (which is a collection of related species) and its unique name within that genus. We belong to the species *Homo sapiens*. There have been other *Homo* species (*erectus, habilis, ergaster*), although all except *sapiens* are now extinct. This two-word convention is an important legacy of the Swedish biologist Carl Linnaeus; the authority for all such names begins with the list in the tenth edition of Linnaeus's *Systema Naturae* (1758). Later descriptions are qualified with the name of the person who identified them. The name *Homo sapiens* occurs in the *Systema Naturae*, and it can therefore be formally referred to as *H. sapiens* Linn (or simply "L").

The black rat was listed by Linnaeus. Its name is *Rattus rattus* L. The brown rat was described in 1769 by an Englishman, John Berkenhout. Its full name is *Rattus norvegicus* Berkenhout. The Pacific or Polynesian rat was originally described in 1848 by Titian Peale, curator of the Philadelphia Museum, as *Mus exulans* from specimens collected by the pioneering United States Exploring Expedition (the Wilkes Expedition) of 1838–1842. Later taxonomists transferred it to the genus *Rattus*; it is now known as *Rattus exulans* Peale. Variation below the species level can be described by a third name (creating a trinomial), which indicates a distinct subspecies or sometimes merely a local form or variant.

A genus may contain only one or a few species, or (as in the case of the fruit fly, *Drosophila*) several hundred. Similar genera are grouped into "families," then "orders," "classes," "phyla," and "kingdoms." *Homo sapiens* is a member of the genus *Homo*; family Hominidae (containing the Great Apes, both living and extinct); order Primates (apes, monkeys, tarsiers, and lemurs); class Mammalia (which incorporates "true" or placental mammals, the pouched marsupials, and monotremes containing the Duck-billed Platypus and a few other mammals which have hair and mammary glands like other mammals but also lay eggs rather than having internal wombs); phylum Chordata (mammals, birds, reptiles, amphibians, fish, and some forms that have a notochord, but not a vertebral column); and kingdom Animalia. All species can be unequivocally labeled in the same way.

Taxonomy is often regarded as a dreary museum-based discipline. This characterization is unfair; taxonomy actually is a testing and imaginative activity. It essentially underpins the whole of biology. A sister discipline is systematics, the study of taxonomic diversity and relationships using structures and, increasingly, the genetic makeups of organisms. Taxonomy, systematics, and ecology have all moved from pure description to increasingly quantitative rigor. At one time, physics and chemistry were described as "exact" sciences, while biology was looked down upon as a rather woolly descriptive science. A more accurate approach now is describing biology and its constituent disciplines as "unrestricted" sciences because they involve a wide range of factors, while the old "exact sciences" are really "restricted sciences," which attained maturity by omitting inconvenient data—such as variation. However one describes ecology, it is an unequivocally quantitative activity, routinely using equations to speak about ecological forces and principles. So we turn to numbers.

Turning Nature into Numbers

In his *Autobiography* Darwin lamented his lack of mathematical capability: "I attempted mathematics but I got on very slowly . . . I have deeply regretted that I did not proceed far enough at least to understand something of the great principles of mathematics; for men thus endowed seem to have an extra sense."

Extra sense or not, mathematics is useful for talking about species and particularly their interactions. Some species are rare, some are very common; some are locally abundant, others occur widely but at low densities. What determines the numbers of a population? The answer to that is given by one of the simplest equations in science:

$$N_{t+1} = N_t + B - D + I - E$$

where the numbers of the population (N) at time (t) increase in a given period by the number of births (B) and immigrants (I), and decrease by the numbers of deaths (D) and emigrants (E). In many situations, we can ignore immigrants and emigrants, and simply count the numbers of births and deaths over a period. If we can do this over the total lifespan of individuals, we can draw a survivorship curve. In some species (like our own), most individuals survive into old age, and then the mortality curve rises steeply. In many insects, marine organisms, and weedy plants, a high proportion of those born die within a comparatively short period, leaving a small fraction to survive and form the parents of the next generation. Other species (many songbirds) have a relatively constant death rate throughout life (see Figure 2.4).

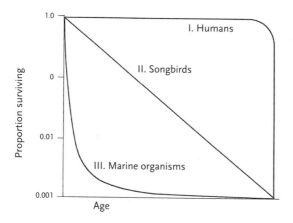

FIGURE 2.4. Different survivorship curves

Key ingredients in population dynamics are fertility and reproductive rates. Some individuals are infertile, such as worker honeybees or ants, but generally speaking we can compute a *reproduction rate*, which is the number of female offspring produced by an average female during her lifetime. In practice, not many organisms realize their full reproductive potential, so a more useful statistic is the *net reproductive rate*, which is the average number of young that an average female produces over her lifetime. Net reproductive rate is calculated by multiplying the fraction of females surviving to each age by the average number of female offspring produced by females at that age, and then adding up the values for the entire life span. This equation can be written algebraically as:

$$R_0 = \Sigma(l_x m_x)$$

where R_0 is the net reproduction rate, l_x is the proportion of females surviving to age x, and m_x is the average number of female young produced by each individual at that age.

Another useful statistic is the intrinsic rate of natural increase or Malthusian parameter, r. It is a measure of the instantaneous rate of change of population size per individual. The maximum values of r range over several orders of magnitude in different species, from about 0.0003 in humans, to 0.015 or so in rats and many insects, 1.0 in the protozoan *Paramecium*, and around 60 in *E. coli*. If r is constant, a population will grow exponentially, but in practice an increase in density is likely to lead to factors that reduce r (e.g., by a shortage of available nutrients or a build-up of a toxin), or even more likely, an increase in mortality.

An intriguing example of the interaction of population growth and limiting factors was a study of a bird, the common swift (*Apus apus*) in the tower of the University Museum in Oxford. Breeding pairs lay two or three eggs. The adults feed their nestlings with insects they catch on the wing. They can almost always catch enough food to nourish two young, so survival in the young of two egg clutches is high. Survival from three egg clutches is lower, although the number of fledglings is greater than from two egg clutches, except in very poor summers when insects are scarce. Consequently it is an advantage for swifts to have a variable clutch size, to make the best of good years while not being penalized in bad years.

Would it help for swifts to lay four eggs or more so that they could overcompensate in good years? This possibility was tested by transferring newly hatched nestlings from one nest to another to make up broods of four. In all four years of this experiment, survival in broods of four was lower than in broods of three, and in two of the years it was even lower than in broods of two (see Table 2.1). Assuming that clutch size is an inherited trait, the implication is that natural selection would act against laying more than three eggs. Speaking of rules or principles in ecology, natural selection would seem to tend to produce median or average forms, weeding out extremes. Selection can be remarkably precise in its effect. Mixing metaphors, we might say that selection functions to "micromanage."

A general correlation exists between reproduction rate and food

TABLE 2.1 Survival in Relation to Brood Size in Swifts, *Apus apus*, in Oxford

Year	Brood size	No. of broods	No. of young dying	Percent lost	No. raised per brood
	2	21	2	5	1.9
1958	3	4	1	8	2.8
	4	2	4	50	2.0
	2	15	0	0	2.0
1959	3	4	0	0	3.0
	4	4	5	31	2.8
	2	18	2	6	1.9
1960	3	6	4	22	2.3
	4	5	14	70	1.2
	2	18	1	3	1.9
1961	3	6	4	22	2.3
	4	5	13	65	1.4

Based on data in Christopher Perrins, "Survival of Young Swifts in Relation to Brood-size," *Nature* 201 [1964]: 1147–48.

supply in mammals, social insects, and certain freshwater copepods. To understand this point, we need to recognize that the actual cause of death of an animal or plant may be quite different from the predisposing cause. For example, when an old person dies of heart failure, the death may be the result of furring of the arteries, deterioration of the kidneys raising the blood pressure, or straightforward inefficiency of heart muscle; when a young person dies from the same "cause," this is more likely to be from a congenital weakness of the heart itself. It is therefore possible to distinguish between the *proximate* cause of death ("heart failure") and its *ultimate* cause (which may be one of a number of mechanisms).

The same argument can be used for natural populations. A starving animal found in midwinter does not necessarily indicate an overall shortage of food. For example, the chance of a wood pigeon surviving its first winter depends almost entirely on its incorporation into a successful flock where it will be better protected against disturbance and protection than if it was by itself.

Another factor affecting survival is stress, which may result from a straightforward lack of food or act *via* behavioral or hormonal mechanisms. The notion of a struggle for existence was important for Darwin, but it did not begin with him or the *Origin of Species*; it was recorded in various ways by observers down the centuries. Three hundred years before Christ, Aristotle commented that "there is enmity between such animals as dwell in the same localities or subsist on the same food. If the means of subsistence runs short, creatures of like kind will fight together." Alfred Tennyson recognized it in his poem *In Memoriam*, completed ten years before the publication of *Origin of Species*: "Nature red in tooth and claw . . . From scarpéd cliff and quarried stone, she [Nature] cries, 'A thousand types are gone: I care for nothing, all shall go.' . . . Are God and Nature then at strife that nature lends such evil dreams?" To understand something about all this, we have to explore the idea of competition both with members of the same species and also of other ones, and how an organism "fits" into its habitat.

ECOLOGICAL NICHES

There is usually an optimal situation for any species: one in which it has both access to sufficient food and protection from its enemies. Just as physical niches in buildings are designed (or are opportunistically available) to contain books, statues, or the like, so there are ecological niches in the natural world that allow the normal functioning of particular individuals. "Niche" is the idea in ecology that seeks to capture this situation by describing the fit between a species and its environment. An ecological niche can be broad or narrow: rats and cockroaches are generalists that can thrive under a wide variety of circumstances; or a niche may be very constrained—as for pandas or many parasites dependent on particular hosts. All niches are bounded by environmental limits for the individual concerned, restricting the distribution of the species to which it belongs.

The original definition of niche effectively envisaged a physical place—a part of a habitat. In fact, though, a niche is not a physical entity in the normal sense. Neither is it simply a species' habitat. It is better described as the role of the species in its community. In his pioneering textbook *Animal Ecology* (1927), Charles Elton (1900–1991) pointed out that the value of the idea of a niche is to provide a way of giving more accurate information than merely stating an animal is a carnivore, herbivore, insectivore, or so on. He defined niche as "the place of an organism in the biotic environment, its relation to food and enemies." Like Darwin, Elton was primarily a naturalist, an observer of the natural world. He wrote that "animals are not always struggling for existence, but when they begin, they spend the greater parts of their lives eating." He wrote,

> The niche of an animal can be defined to a large extent by its size and food habits.... [For example, there] is in every typical community a series of herbivores ranging

from small ones (e.g., aphids) to large ones (e.g., deer). Within the herbivores of any one size there may be further differentiation according to food habits. . . . We might take as a niche all the carnivores which prey upon small mammals and distinguish them from those which prey upon insects. . . . [However] it is convenient sometimes to include other factors than food when describing the niche of any animal.

As in the debate over defining "ecosystem," which pitted a hypothetical against a more empirical view, the concept of the niche has also been buffeted in various ways. Is the niche an ideal state or a messy, changing state of affairs in nature? One of Elton's contemporaries, the Yale ecologist Evelyn Hutchinson (1903–1991), expanded Elton's original interpretation. Since a niche is the sum of all the environmental factors acting on an organism, Hutchinson defined it somewhat gnomically as "a region of an n-dimensional hyperspace, comparable to the phase-space of statistical mechanics." The idea of n-dimensionality simply reflects the fact that a species can only survive and reproduce within an unspecified and possibly large range of bounds: temperature, humidity, food availability, and so on. But other species almost certainly affect these limits. It is easy to envisage situations where there is ample food (i.e., an energy source) available but an individual is prevented from using it for one reason or another (such as the presence of a predator). The *fundamental* niche of a species—its supposedly *ideal* niche—is the range it would occupy in the absence of interference from other species. In practice, a species is likely to be confined by competitors or predators to a much more restricted range; this is its *realized* niche.

The idea of a fundamental niche has practical problems:

▸ Since it has an infinite number of dimensions, we cannot actually determine the niche of any organism.

▸ It assumes that all environmental variables are linear and can be measured.

▸ The model refers to a single instant in time, although interactions in reality are dynamic.

The most widely used concept nowadays is the *resource-utilization* niche, formulated by Princeton ecologist Robert MacArthur (1930–1972) and his Harvard collaborator Richard Levins (1930–). Like the Elton-Hutchinson model, the resource-utilization niche is quantitative and multidimensional, but it focuses on what members of a species population actually do—in particular, how they use resources. This approach frees the niche from the difficulties of Hutchinson's model; indeed, it is really little more than a precise description of the natural history of a species: its habitat, food types, activity times, and so on. The MacArthur-Levins understanding is effectively a refinement of Elton's approach, based on the same principles of accurate observation rather than abstract theorizing.

All this brings us back to the bottom line: the non-negotiable requirement every living organism has for energy. All living processes need energy. The ultimate source of energy is, of course, the Sun. The total amount of solar energy striking the Earth's surface every day is equivalent to the energy in 684 billion tons of coal—sufficient to melt a ten-meter (thirty-foot) thick layer of ice over the whole surface of the Earth. Around 98 percent of this energy is reflected or absorbed, and only half of the remaining 2 percent is in a wavelength that plants can utilize for photosynthesis.

Ultimately, this is the energy for life. Plants "fix" energy received from sunlight, although for this to happen they also have to absorb water and carbon dioxide and are constrained by temperature and soil, plus a supply of such nutrients as phosphorus and organic nitrogen, and trace elements like magnesium. The process does not stop even there: plants have to apportion their energy use between growth, protection, and reproduction. Animals get their energy from eating plants or other animals, and have the same problem of dividing their available resources between growth and reproduction. All these factors determine where an animal or a plant finds its place or "niche" in the community. This is obviously much more than its position in space. If we describe its physical place as its habitat—effectively its home address—we can then regard its niche as its profession, probably sharing it with other species.

Fifty years ago, the study of our energy resource relationships seemed an attractive way to produce simplifying generalizations. The International Geophysical Year (1957–1958) showed the benefit of coordinating research on a global scale. It was work from this period that confirmed the existence and role of tectonic plates. Could ecologists profit from a similar enterprise? Beginning in 1964 a decade-long International Biological Program (IBP) invested enormous resources to seek generalizations about energy input and use in major ecological systems. But by focusing on the big picture—the energy flow in whole biomes—little was learned about basic and local ecological interactions. Vast amounts of data were collected and complex energy models developed, but they did not lead to any useful generalizations. The IBP certainly raised the level of international collaboration among ecologists, but its most important legacy seems likely to be the recognition of the value of long-term data sets for detecting the effects of changed environmental conditions.

A number of such long-term studies have continued over many generations: some agricultural plots established at the Rothamsted Research Station in Britain have been monitored since the mid-nineteenth century, flying insects have been collected in suction traps at Rothamsted since 1931, plankton in the North Atlantic have been surveyed continuously since the same date. A major study of ecosystem chemistry has been running at Hubbard Brook in New Hampshire since 1963 and has provided data to test various hypotheses about energy flow in ecosystems (and, incidentally, proved the existence of "acid rain" far from the industrial plants where it originated). Concern about climate change has stimulated a search for historical records. We know, for example, that plants are now flowering a week earlier around Walden Pond than in the 1850s when Thoreau was recording there.

A similar result stretching back to 1760 has been obtained for more than four hundred flowering plant species in Britain. Much more importantly, measurements since 1958 of carbon dioxide in the atmosphere above the extinct Hawaiian volcano of Mauna Loa have shown the continuing rise of CO_2 concentration in an area as remote as possible from polluting sources; these data have been a significant factor in establishing the reality of climatic change.

The lack of any significant conclusions from the IBP studies recalls the concern of another of the ecological greats, a Russian, Georgii Gause (1910–1986). Gause believed that the complexity of niches in the natural world was too great for useful study. He concentrated instead on laboratory experiments between two single-celled organisms (*Paramecium*) and a predator, another ciliate (*Didinium*). He grew cultures of *Paramecium* and then introduced the predator at various times or in various numbers. Always the prey became extinct, inevitably followed by the predator, which had no food. But if Gause provided a refuge for the *Paramecium*, the species could survive. He proposed a Principle of Competitive Exclusion (often known as Gause's Law), that two species competing for the same resources cannot stably coexist if other ecological factors are constant. One of the two competitors will always overcome the other, leading to either the extinction of the competitor or a shift toward a different ecological niche. This result has been confirmed many times.

To take just one example, Tom Park in Chicago cultivated two species of flour beetles, *Tribolium confusum* and *T. castaneum*. Either growing by itself survived a wide range of environmental variation; if grown together, one or the other of the species always did better under test conditions (Table 2.2). If the environment was diversified by putting wheat husks or pieces of glass tubing into the culture, both species survived indefinitely. In other words, a diverse habitat provided conditions that allowed species diversity to exist and almost certainly mirrored the natural habitat of the beetles more accurately than pure flour.

TABLE 2.2 Competition between Flour Beetles,
***Tribolium confusum* and *T.castaneum* in Experimental Climates**

(based on T. Park, "Experimental Studies of Interspecies Competition
II. Temperature, Humidity and Competition in Two Species of Tribolium,"
Physiological Zoology 27 [1954]: 177–238)

	Percentage wins	
Climate	*T. confusum*	*T. castaneum*
Hot-moist	0	100
Temperate-moist	14	86
Cold-moist	71	29
Hot-dry	90	10
Temperate-dry	87	13
Cold-dry	100	0

In experiments with the two species, one species was always eliminated, but the balance between the species changed with the conditions. In intermediate climates, the outcome was probable rather than predetermined; even the inferior competitor occasionally achieved a density enabling it to out-compete the other species.

Such results are clear-cut in laboratory conditions and may help to understand what happens under natural situations. A rather spectacular example is the predator-prey cycles of some species in Arctic regions. Here there are periodic irruptions of lemmings and snowshoe hares followed a year or two later by outbreaks of lynxes and other predators that feast on the hapless lemmings or hares. The result is that the prey becomes much rarer, and the predators suffer from food shortages, becoming much less common in turn (see Figures 2.5 and 2.6). To generalize from such a situation, the problem is recognizing and evaluating different effects:

▶ Competition must not only be likely, but it should also have detectable effects on survivorship, fecundity, or some other trait.

▶ The resources for which the organisms are competing must be in short supply.

▶ Competition may be density-dependent—that is, the probability of an individual being affected increases with the number of competitors.

▶ There needs to be some sort of reciprocity.

The last consideration is probably the most confusing in the interaction scenario. Species may cooperate as well as compete.

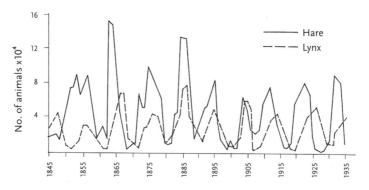

FIGURE 2.5. Changes in abundance of lynx and snowshoe hares as indicated by pelts received by the Hudson's Bay Company. Note that the numbers of lynx tend to peak just after the rise in numbers of hares (after David A. MacLulich, "Fluctuations in the Numbers of the Varying Hare [*Lepus americanus*]," *University of Toronto Studies, Biology Series* 43 [1937]: 1–136, by permission of University of Toronto Press; and Charles S. Elton and M. Nicholson, "The Ten-year Cycle in Numbers of the Lynx in Canada," *Journal of Animal Ecology* 11 [1942]: 215–44, by permission of Wiley-Blackwell).

FIGURE 2.6. (opposite) Food web in the boreal forest of North America, showing the importance of lynx as a predator and snowshoe hares as prey (after Nils Christian Stenseth et al., "Population Regulation in Snowshoe Hare and Canadian Lynx: Asymmetrical Food Web Configurations between Hare and Lynx," *Proceedings of the National Academy of Sciences of the USA* 94 [1997]: 5147–52, copyright 1997, National Academy of Sciences, U.S.A.)

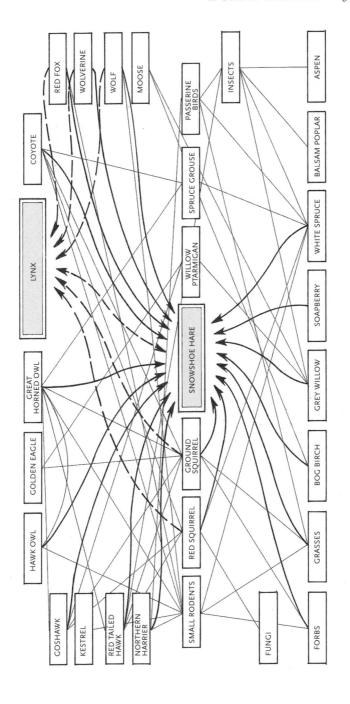

Lichens are composites of a fungus and an alga or bacterium. Many trees grow better when their roots are in association with mycorrhyzal fungi. Clown fish and sea anemones may protect each other by living close together. Some ant species "cultivate" aphids. There are many such examples of mutual benefit and an extensive vocabulary has grown up of the ways in which *symbiosis* may take place (commensalisms, mutualism, etc.). Sometimes the benefit is only to one partner (*parasitism*), and sometimes one partner may be damaged or even killed (*predation*).

Some of the most spectacular examples of species interactions are when a new disease attacks a susceptible population. HIV-AIDS has caused hundreds of thousands of deaths in the last few decades since it crossed the boundary between apes and humans, but it is only the most recent epidemic disease affecting human populations. In the fourteenth century, the Black Death (almost certainly bubonic plague, transmitted by rat fleas) killed almost half the population of Europe. Smallpox and measles were major scourges of hitherto unexposed native peoples when introduced from Europe into North America and the Pacific. John Snow mapped the cases of cholera in the London district of Soho and stopped the local epidemic by removing the handle of a pump which was the source of water infected with the cholera germ for the surrounding inhabitants (see Figure 2.7). The study of such situations is known as epidemiology; its principles are derived wholly from a consideration of the interaction of species in basic ecology.

At this point, it is probably helpful to go back to the basic population equation:

$$N_{t+1} = N_t + B - D + I - E$$

Another way of writing this is:

$$dN/dt = B + I - D - E$$

where dN/dt is the calculus shorthand for the rate of increase of numbers (N) with time (t). In a closed population with no immi-

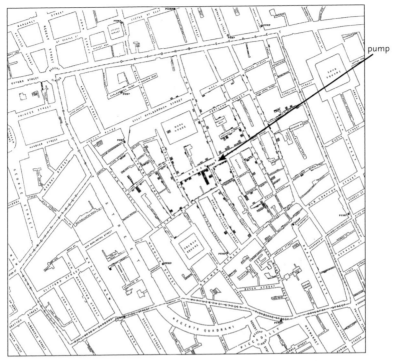

pump

FIGURE 2.7. Map of the distribution of cases of cholera in Soho, London, in 1854, prepared by John Snow. Contrary to the belief at the time, Snow was convinced that the disease was waterborne. On the basis of the occurrence of cases, Snow was persuaded that the source of infection was a particular communal pump. He made the pump inoperative by removing its handle. Additional cases of the disease were very few.

gration or emigration, the absolute number of births or deaths in a given interval of time depend on the number of individuals in the population. Moreover, if we assume average fertilities and death probabilities, B will equal bN and D will equal dN.

We can then rewrite

$$dN/dt = B - D \text{ as } dN/dt = bN - dN,$$

or

$$dN/dt = (b - d)N$$

This is known as the *logistic equation*. If the numbers of births and deaths are independent of each other, $(b - d) = r$, which is the *intrinsic rate of increase* of the population. If b is greater than d, the population will increase without limit, or *exponentially* in mathematical language.

The assumption that b and d are independent is almost certainly false in most situations. As a population increases, the amount of food and space available to each individual decrease, resulting in a reduced survival rate (which has the same effect: increased mortality). There may well also be a reduction in the birth rate, except in widely spaced small populations. But here, individuals may be so far apart that they have difficulty finding a mate.

If, however, we make the common and reasonable assumption that the increase and decrease are linear, we can use the standard equation of one variable (y) on another (x),

$$y = a + bx$$

where a is the intercept of the line on the y axis when $x = 0$. We can then state the dependence of b and d on N as

$$b = b_0 - k_b N \text{ and } d = d_0 + k_d N.$$

In this case, b_0 and d_0 are the values approached as the population becomes very small, k_b is the slope of decrease for the birth rate, and k_d is the slope of increase for the death rate. Substituting these values into the logistic equation, this becomes

$$dN/dt = [(b_0 - k_b N) - (d_0 + k_d N)]N$$

When births and deaths equal each other, the population size is stable—that is, when

$$b = d \text{ or } b_0 - k_b N = d_0 + k_d N,$$

then

$$N = b_0 - d_0/k_b + k_d$$

This value of N is called the *carrying capacity of the environment,* usually denoted as K. This relates to an equilibrium population size that will be approached in time from any initial size: if N is greater than K, the population will decrease in size; if N is less than K, it will increase (see Figure 2.8).

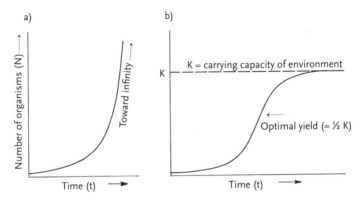

FIGURE 2.8. Population growth curves. a) Exponential growth. b) Logistic growth.

One final rearrangement:

since $K = b_0 - d_0/k_b + k_d$
and the intrinsic rate of increase, $r = b_0 - d_0$,
the logistic equation can be rewritten as
$$dN/dt = rN([K - n]/K).$$

In this form, the equation is often called the *Lotka-Volterra equation*, after the researchers who proposed it in the 1920s. It provides the quantitative basis for many of the fundamental interactions in ecology, such as the predator-prey relationship already mentioned (Figure 2.5). An intriguing and unexpected result is that if two interacting species are destroyed at the same rate, such as by indiscriminate hunting or pesticide use, the prey will increase disproportionately as the predators will decrease. This effect has important practical implications, explaining, for example, why insect pests treated with an insecticide often recover to higher levels than previously (see Figure 2.9). The reason, of course, is that natural predators and parasites of the pest species are also destroyed by the pesticide treatment.

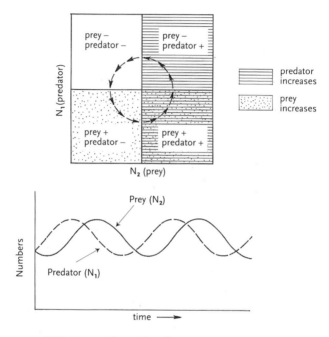

FIGURE 2.9. Effect on numbers of predator-prey interactions as predicted by the Lotka-Volterra equation. Prey numbers increase when predator numbers are low, leading to an increase in predator numbers and driving down prey numbers. This leads to an oscillation in numbers (see lower figure). Compare this with the lynx-snowshoe hare interaction (Fig. 2.6).

The same theory can be employed to calculate the maximum sustainable yield of a resource. It has been most commonly used in computing the potential yield of marine organisms—how much fish can be caught without depleting the overall stock. In other words, it shows what is the maximum sustainable harvest, an important fact from a commercial point of view. This balancing act is not easy, and the calculations are often ignored. The most sensible strategy for a harvester is to take as much as possible from a population without overexploitation; the harvester's behavior can be regarded as operating in the same way as a predator or some other sort of competitor. When an equilibrium population is harvested, numbers will fall and recruitment increase. If too high a proportion is removed (that is, when deaths heavily outweigh recruitment), the population may not be able to maintain viable numbers. The optimum removal rate is when the recruitment rate is highest—at the steepest part of the curve of numbers, where harvesting and recruitment rates are equal and opposite. It represents the maximum sustainable yield of the resource and is at $K/2$ for populations whose dynamics are describable by the logistic equation.

This calculation is oversimplistic in that it treats all individuals and all environments alike, whereas different age classes and heterogeneous habitats may affect the population growth curve markedly. Notwithstanding these caveats, such calculations are commonly used as the basis for sustainable commercial fishing and hunting recommendations.

Food Webs

The real life of animals is a compound of many things: fixed
and predetermined limits impressed by the environment;
the relations of the sexes; the survival of things that are useful;
a certain free will in the matter of choosing between good and evil
surroundings, accompanied by a great deal of movement; a fair
amount of pure chance; and sometimes a growing stock of new
ideas born out of contact with new situations—predetermination,
sex, materialism, free will, destiny, originality and tradition.
The study of limiting factors in the surroundings of animals
enables us to define the stage and the scenery in the midst of
which the act will be played. But the actors, starting though they
do with fixed instructions and fixed limits of time, are still partially
responsible for making the play a success or a disaster.
—Charles Elton

Much of this detail about population interactions can be brought together in the form of a food web or chain. A food web is a way in showing the transfer of energy through a variety of organisms from its source in plants which fix the energy from the Sun. A food web can be described in terms of *trophic levels* (that is, nutritional levels), each level being a part of a web in which a group of organisms obtains food in the same general way. Thus, all animals that get their energy from eating grass—grasshoppers, voles, cattle—can be regarded as part of the same trophic level. A typical ecosystem has between three and six trophic levels—also called "food links"—through which energy passes. In aquatic systems, algae, phytoplankton, and other aquatic plants occupy the same trophic level as grass. Here, small crustacea, some insect larvae, and a few herbivorous fish occupy a similar trophic level as cattle.

Some energy is lost at each level, so a short food chain leads to more living matter (*biomass*) surviving than in a longer one. A five-link food chain (such as alga-crustacean-insect-minnow-bass) is considerably less energy-efficient than a three-link chain (such as alga-minnow-bass). Antarctic seas tend to have short, simple food chains (alga-shrimp–baleen whale at the simplest) and as a result are among the most productive oceans in the world. The primary herbivore here is that of a small shrimp, *Euphausia superba* (a species of krill). During the polar summer these animals have a twenty-four-hour energy input from sunlight and an upwelling of nutrients favoring plankton growth. Krill is the primary food source of fish, birds such as penguins, and baleen whales. The major predators at the top of the trophic structure are seals, squid, sperm whales, small toothed whales, and, since the middle of the eighteenth century, humankind.

Trophic structures tend to be more complex in temperate and equatorial regions. Figure 2.10a shows some of the more important food links in a freshwater pond. It illustrates the interaction between simple chains and their weaving into a "web." In most temperate food webs, the pattern of energy flow becomes so complicated that it is difficult to envisage all the relationships. For example, in a typical temperate wood, forty or fifty species of insectivorous birds may be feeding on several hundred of species of insects, each of which has its own suite of parasites.

The more complicated a food web, the more stable it will be. If one species declines or disappears, there are likely to be other sources of food for species that depended on it. The converse of this is the fragility of many agricultural systems. Agriculture is really the applied management of food chains, increasing efficiency and productivity, but at the expense of ecological vulnerability and instability. Consequently a novel attack on a crop may have catastrophic results (e.g., potato blight in Ireland in the 1840s, *Phylloxera* in grapevines in France at the end of the nineteenth century, killer honeybees spreading in the United States since the mid-twentieth century).

Another feature of energy flow in food chains is that they give rise to trophic pyramids. The loss of energy at each level means that progressively less is available for higher-level predators. There are less food opportunities for, say, a lynx or an eagle or a tiger than for a deer or a vole. As a result, top predators have to scavenge over a wider area than a generalized herbivore.

These considerations help us to bring together a number of the ideas described in this chapter by recognizing two basic types of ecological strategies, which we can call "opportunist" and "equilibrium." As the names suggest, species using the former strategy respond quickly and prodigally to any opportunity, while equilibrium species fluctuate relatively little in number in good times and bad. The distinction between opportunists and equilibrium species is somewhat artificial, because the two sorts are really the extremes of a continuum of ecological strategy, but it captures a real effect in the natural world.

Populations of annual plants and insects typically grow rapidly during spring and summer but decline steeply with the onset of cold weather. In contrast there are other species that exist at much

a)

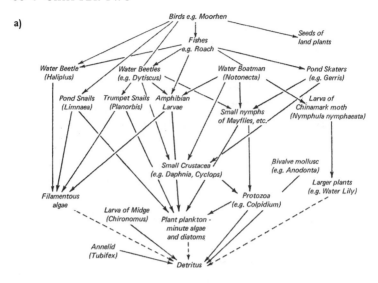

b)

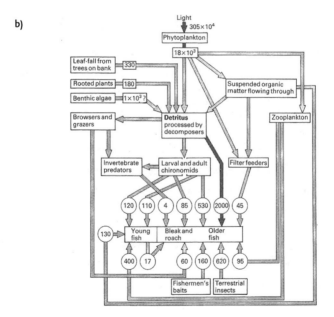

FIGURE 2.10. a) Simplified food chains in a freshwater pond (after W. H. Dowdeswell, *Animal Ecology* [London: Methuen, 1961]). b) Energy flow in kilojoules (kJ) per square meter per year in the River Thames at Reading, measured by determining the amount of sunlight energy received in a given

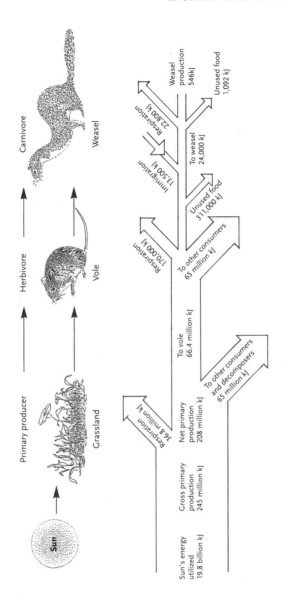

Primary producer — Herbivore — Carnivore

Sun — Grassland — Vole — Weasel

Sun's energy utilized 19.8 billion kJ

Gross primary production 245 million kJ

Respiration 36.8 million kJ

Net primary production 208 million kJ

To vole 66.4 million kJ

To other consumers and decomposers 65 million kJ

Respiration 170,000 kJ

To weasel 24,000 kJ

Unused food 311,000 kJ

To other consumers 65 million kJ

Immigration 13,500 kJ

Respiration 22,800 kJ

Weasel production 546kJ

Unused food 1,092 kJ

time and the energy value of the tissues of the species in the ecosystem (from W. H. Dowdeswell, *Ecology: Principles and Practice* [London: Heinemann Educational, 1984]; based on data collected by Alasdair Berrie). c) Energy loss (in kJ) between trophic levels in a grassland food-chain.

more stable densities, provided that their resources do not vary greatly. Most vertebrates fall into the equilibrium group. They are normally in equilibrium with their resources.

Opportunistic populations tend to be density independent in their regulation and have a high reproductive rate (r). In contrast, equilibrium species are strongly density dependent. Their numbers stay close to the environment's carrying capacity (K). The distinction is between species that are highly mobile and good colonizers, but have a low competitive ability, so their populations tend to be ephemeral; and others that put their energy into maintenance and resisting competition. Such traits are clearly inherited. The two sorts of strategies can be conveniently labeled r and K, produced by r-selection and K-selection, respectively.

In general, r-strategists (opportunists) live in variable or unpredictable environments and are subject to recurring heavy mortality; selection favors rapid development and early reproduction, high reproductive rates, and low competitive ability. In contrast, selection for a K-strategist (equilibrium type) leads to slower development, a greater body size, and the ability to withstand both intra- and inter-specific competition. The strategies are not really alternatives: species range in their survival stratagems as widely as the stability of the habitats they occupy (Table 2.3).

All these factors relate to the niche of particular species. Humans, mice, and cockroaches may live in the same area, but they have totally different experiences of the environment, largely as the result of their sizes and abilities to move; obviously a human rides roughshod over many features that are major obstacles to a mouse. One way of looking at this is to think of an environment as having "roughness" or "grain". Small, relatively immobile creatures tend to experience environmental factors as sets of alternatives. Their environment is "coarse-grained" and even if the overall environment for the species is constant, it will be uncertain for an individual. With greater mobility, environmental differences become increasingly fine-grained and are experienced as a succession of differences with

TABLE 2.3 Correlates of *r* and *K* Selection

	r selection	*K* selection
Climate	Variable and/or unpredictable; uncertain	Fairly constant and/or predictable; more certain
Mortality	Often catastrophic, nondirected, density independent	More directed, density dependent
Survivorship (see Figure 2.4, p. 40)	Often Type III	Usually Types I and II
Population size	Variable in time, nonequilibrium; usually well below carrying capacity of environment; unsaturated communities or portions thereof; ecologic vacuums; recolonization each year	Fairly constant in time, equilibrium, at or near carrying capacity of the environment; saturated communities; no recolonization necessary
Intra- and interspecific competition	Variable, often unimportant	Usually keen
Selection favors	Rapid development	Slower development
	High maximal rate of increase, r_{max}	Greater competitive ability
	Early reproduction	Delayed reproduction
	Small body size	Larger body size
	Single reproduction	Repeated reproduction
	Many small offspring	Fewer larger progeny
Length of life	Short, usually less than a year	Longer, usually more than a year
Leads to	Productivity	Efficiency
Stage in succession	Early	Late, climax

Based on Eric Pianka, "On *r*- and *K*-selection," *American Naturalist* 106 (1970): 592–97.

a similar average for all members of a population. Their environment is therefore more certain. Vertebrates with generally larger
body size, greater mobility, and better homeostatic control than
invertebrates need less ability to cope with environmental alternatives. They tend to be K-strategists.

The idea of grain includes not only the sizes of patches relative
to the size and mobility of the organism but also variations in time
and space of temperature, food, parasites, and so on. Environmental grain may be different at different stages of the life cycle. Lar-
val insects have a coarser-grained environment than winged adults.
Social dominance may affect grain, given that a dominant animal is
less restricted than a subordinate one.

One implication of being an r-strategist (an opportunist type)
is that some populations rarely, if ever, come into equilibrium with
their food supply. Habitat, niche, and survival strategy are ecological ideas that overlap. They are also helpful for understanding genetic processes, since all the relevant environmental factors
act together to affect survival and reproduction and hence natural
selection. These considerations help us to reach some general conclusions about the genetic processes acting in natural situations:

1. It is wrong and frequently misleading to speak of the strength
of natural selection acting on a population or species, because
selection depends on the changing nature of the habitat or niche in
either or both time and space. This implies that a population may
benefit from being genetically variable; as soon as one adapted type
increases in frequency in response to an environmental constraint,
the environment may change (either with time or with the movement of the organism). In general, organisms living at depth in the
sea, where the environment is relatively constant, tend to have less
variation than their surface-living relatives.

2. The recognition of different environmental strategies helps
to clarify the importance of different types of selection acting at
different times during the life cycle. Intriguingly, experiments to

understand such effects have shown that some genetic changes (mutations) can be beneficial. For example, radiation-induced mutations are not unremittingly deleterious; some have been shown to increase the competitive ability (i.e., "niche-width") in one species of *Drosophila* when cultured in the presence of another.

3. Some arguments between ecologists (such as the control of reproductive rates in different species) can be resolved by taking into account the conditions under which the species concerned evolved. For example, an oft-repeated claim that density dependence is unimportant in controlling population size is based almost entirely on studies of opportunist r-type species (especially insect pests), which colonize empty habitats and rarely reach equilibrium numbers.

Opportunism is patently pragmatic and only partly predictable. In contrast, K-strategies can be caricatured as nothing more than the consequences of optimization. Optimization has become recognized in recent years as one of the most important ordering forces in ecology. The concept originated from the development of control mechanisms in engineering. It involves the choice of the best overall tactic when alternative courses are available, that is, the way an organism trades off the advantages against the disadvantages of its possible "investments." We can usefully use the concept of "fitness," an idea that in biology is a measure of reproductive success and not a description of physical prowess. For any organism, we can identify five key areas that contribute to its "fitness":

1. Extent of investment in physiological adaptation to inclement physical conditions. For example, the amount of fat a bird accumulates in the autumn is a trade-off between protection from cold and the loss of agility when faced with a predator's attack; similarly, the energy resources necessary for the success of long-distance migration have the penalty of having to be carried.

2. Extent of investment in defense to avoid death by predation or, for plants, reduced fitness by herbivory—such as heavy armoring (e.g., thorns) or poisonous metabolites.
3. Investment in food harvesting and development of the particular traits—such as storage organs in plants.
4. Investment in reproductive activities—finding a mate, successful breeding, care of offspring: Is it better to produce a few large offspring (as do elephants) or many small ones (such as herrings or aphids)?
5. Investment in escape—such as migration or dormancy.

These tactics—or investments—can be combined into a reproductive success matrix that brings together virtually all the ecological properties and challenges set out in this chapter (Figure 2.11a–c). "Nature" is almost impossibly complex and intertwined, but it is made up of elements that we can begin to understand and study. Such analysis is the core of ecology. The danger is that when we think we have understood some important process—photosynthesis, say, or energy flow or environmental instability or gene action—we think we have gotten to the bottom of ecology. Sadly that is not so. Ecology does not yet have its Newton or Einstein,

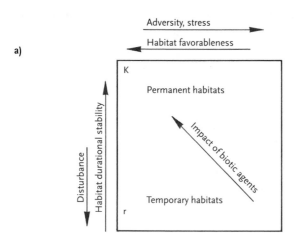

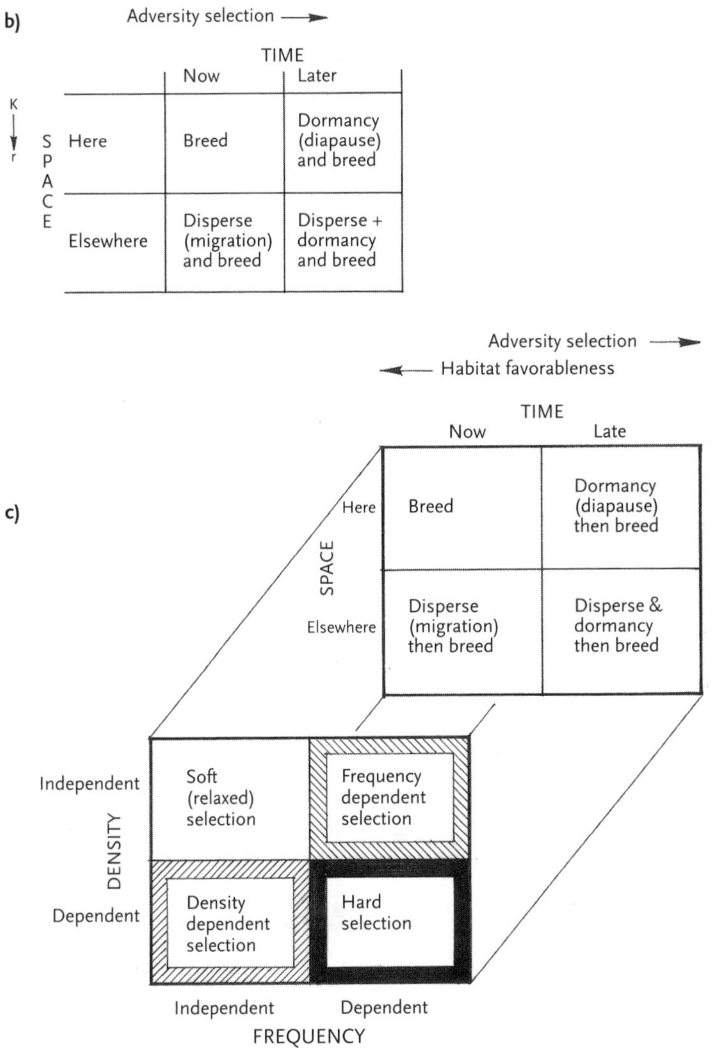

FIGURE 2.11. a) (opposite) Habitat type and biological response (from T. R. E. Southwood, "Habitat, the Templet for Ecological Strategies," *Journal of Animal Ecology* 46 [1977]: 337–65, by permission of Wiley-Blackwell). b) Different tactics in different habitats (after T. R. E. Southwood, "Tactics, Strategies and Templets," *Oikos* 52 [1988]: 3–18). c) Different forms of natural selection as they relate to the habitat templet (from R. J. Berry, "Ecology: Where Genes and Geography Meet," *Journal of Animal Ecology* 58 [1989]: 733–59, by permission of Wiley-Blackwell).

although the subject is now reaching a maturity that a unifying synthesis may be not too far away.

One possible contender for an ecological synthesis is the Gaia theory (see p. 107); another more orthodox—but still contentious—idea is the neutral theory of biodiversity put forward by the tropical botanist Stephen Hubbell (1942–). Contrary to the usual assumption of communities composed of different niches maintained by internal interactions (and even further away from Clements's picture of mutually integrated super-organisms), Hubbell argues that individuals of every species at a given trophic level in a food web reproduce and die independently of other species in the community; their coexistence is neutral. In its original form, Hubbell's theory is probably too all-encompassing, but it may well show a possible way forward.

We are a long way from the end of our ecological journey. Any final synthesis will have to acknowledge Lovelock's Gaia and Hubbell's neutral theory, and will have to join the work and insights of ecologists proper with the legacy of evolutionists like Charles Darwin and geneticists like James Watson and Francis Crick (who sparked the molecular revolution with their structure of the DNA molecule). And an understanding of ecology will also have to take on board the observations and sheer wonder passed down to us by such people as Gilbert White, Joseph Hooker, John Muir, David Attenborough—and many others. Ecology is an exciting subject at an exciting phase in its development.

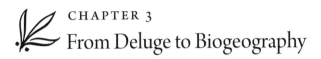

CHAPTER 3

From Deluge to Biogeography

We never know the worth of water until
the well is dry. —Thomas Fuller

THROUGHOUT MOST of human history, we had no reason to
doubt that we were living in a world unchanged from its beginning, and according to Plato writing four centuries before Christ, a
world that is virtually unchangeable. From this belief to an assumption that everything exists for the benefit of human beings is only
a short step, involving a belief natural enough in itself and seemingly in accord with the Bible. This meant that for many centuries
there seemed to be no need to study the natural world for its own
sake; knowledge about it came either from the information necessary to survive or from philosophical speculation about the ways
of the Creator.

The brake on our perception is well illustrated by the comment
of Harvard biologist Ernst Mayr:

> Plato's dogmas had a particularly deleterious impact
> on biology through 2000 years. One was essentialism,
> the belief in a constant *eide* separate from and indepen-
> dent of the phenomena of appearance. The second was
> the concept of an animate cosmos, a living harmoni-
> ous whole, which made it so difficult in later periods to
> explain how evolution could have taken place, because

any change would disturb the harmony. Third, he [Plato] replaced spontaneous generation by a creative power. A demiurge. And fourthly [was] his great stress on "soul."

This Platonic constraint began to lessen in the seventeenth century; indeed by its end, discoveries from astronomical observations and geological processes led to ideas of changelessness becoming increasingly untenable. The existence of deep time and deep space was widely accepted. This shift in perception was necessary for science—never mind ecology—to develop. It was not so much a denial of our heritage as a reinterpretation of the information on which it was based in the light of new discoveries. We began to recognize that the world is not here solely for our benefit and that it is bigger, grander, and older than our forebears ever suspected.

A key actor in this story was an English naturalist, John Ray (1627–1705), sometimes called "the father of natural history." Ray argued that nature can be studied for its own sake, on its own terms, and that it exists for reasons of its own, not just to serve humans. He was certainly not a natural iconoclast. He was the son of a blacksmith, born in a small village in eastern England, but he seized the opportunities available to him and improved the traditions he inherited, particularly in exercising his faculty for observation and description. He lived in a time of social and intellectual turbulence—the flowering of the Age of Reason, a period that included Galileo's debates in Rome as well as the philosophical challenges of John Locke and David Hume. During Ray's lifetime, agricultural improvements were gathering pace; when he was middle-aged, the rapidly growing London was almost destroyed in a Great Fire (1666) and then rebuilt under the genius of Christopher Wren. He was an early fellow of the Royal Society of London, being admitted in 1667, seven years after the Society's founding; he shared in the adolescence of the reformed Church of England, resigning his fellowship of Trinity College, Cambridge (where Isaac Newton was one of his contemporaries) because of his reservations about the

1660 Act of Uniformity, which required all citizens to adhere to an official creed.

During the last quarter of the seventeenth century, Ray published under his own name or that of his pupil and patron Francis Willughby, classifications of the major groups of animals and plants, taking a major step toward a natural system and preparing the way for Linnaeus, both in terms of method and of data. The French biologist Baron Cuvier described Ray as "the first true systematist of the animal kingdom . . . His works are the basis of all modern zoology." Gilbert White of Selborne regarded Ray as his mentor, scientifically and theologically. He extolled Ray as "the only describer that conveys some precise idea in every term or word, maintaining his superiority over his followers and imitators, in spite of the advantage of fresh discoveries." More soberly but perhaps more importantly in recognizing his legacy was the clarification in his book *Ornithology* (1678) that "what properly relates to Natural History," involves "wholly omitting Hieroglyphics, Emblems, Morals, Fables, Presages or ought else appertaining to Divinity, Ethics, Grammar, or any sort of Human Learning." By separating the study of nature from speculative religious concerns, Ray liberated the study of the natural world from the specifically religious constraints of biblical interpretation and opened the way for ecology; Ray followed the Protestant Reformers in seeing the book of scripture as free of allegory, to be interpreted only in a literal and historical sense. Put another way, Ray applied the implications of the belief—common in his time—that God had written a book of nature as well as a book of scripture. This belief in the two separate books showed the absurdity of using religious allegory to explain nature. The book of nature stood on its own. He declared in his book *The Wisdom of God Manifested in the Works of Creation* (1691):

> It is a generally received opinion that all this visible world was created for Man [and] that Man is the end of creation, as if there were no other end of any creature

but some way or other to be serviceable to Man. . . . But
though this be vulgarly received, yet wise men nowadays
think otherwise.

Ray's statement provided a motive for the study of the natural
world for its own sake. He also challenged the dominant assumption of his age that, under God, all things are subordinate to human
beings

The decades after Ray's death saw the use of reason to explore
the nature of God reach a climax. The beginning of the eighteenth
century was the heyday of "physico-theology," epitomized by the
Boyle Lectures, founded with a bequest from the chemist Robert
Boyle (1627–1691) to be delivered in London churches and directed
against unbelievers. A particularly notable series, given in 1711–12 by
William Derham, was titled *Physico-theology or a Demonstration of
the Being and Attributes of God from His Works of Creation*. Derham
based his arguments on Ray's *Wisdom*, acknowledging "my Friend,
the late great Mr. Ray."

Physico-theology treated God as the First Cause, the Divine
Watchmaker who had created all things and then retired above
the bright blue sky. It reached its climax in the writings of Archdeacon William Paley (1743–1805), particularly his *Natural Theology* (1802), which reused Ray's examples in the *Wisdom* without
acknowledgment. Although the two books have much in common,
they are poles apart: Ray worshipped God for his marvelous works
in nature (theism); Paley saw nature as the proof of God's handiwork (deism). The last major manifestation of this traditional version of natural theology were the Bridgewater Treatises, endowed
by the Reverend Francis Egerton, eighth Earl of Bridgewater, who
(allegedly in expiation for a misspent life) charged his executors,
the archbishop of Canterbury, the bishop of London, and the president of the Royal Society of London to pay eight scientists one
thousand pounds each to examine "the Power, Wisdom, and Goodness of God, as manifested in the creation; illustrating such work

by all reasonable arguments." Their books were published between 1833 and 1837.

Interest in natural theology—extrapolating theology from science—had social benefits, turning the attention of ordinary people to a study of the natural world. The mid-nineteenth century saw crazes for drawing-room aquaria and for growing exotic ferns. A survey in 1873 revealed 169 local scientific societies in Britain, of which 104 claimed to be field clubs. Most of these had come into being since 1850, a rate of formation of 10 per year. More and more people were studying the natural world—not as hunters or farmers but as enquirers.

But knowledge of the world was moving on; traditional natural theology was dying. The early geologists were providing more and more evidence of a long Earth history. Alongside their idea of "deep time," it was becoming apparent that the universe was vastly bigger than envisaged by mediaeval astronomers. The telescopes of William Herschel (1738–1822) and others showed that the solar system itself was moving through space. There is "deep space" as well as "deep time." At the same time that Chalmers, Kidd, Whewell, Bell, Roget, Buckland, Kirby and Prout were laboring over their Bridgewater Treatises, Charles Darwin was sailing round the world in HMS *Beagle* (1831–36). The outline of his evolutionary ideas were in place by early summer 1837. Then in 1844 Robert Chambers's *Vestiges of the Natural History of Creation* was published. It was effectively a tract against Paley's version of deism. The *Vestiges* was an immediate best-seller. In its first ten years it sold more copies than Darwin's the *Origin of Species* did fifteen years later, yet it was full of errors. For Darwin, "The prose was perfect, but the geology strikes me as bad and his zoology far worse." Nevertheless it stirred debate. Darwin welcomed it on the grounds that "it has done excellent service in calling in this country attention to the subject and in removing prejudices."

Chambers wrote that when there is a choice between God's direct "special creation" of things in nature and the operation of general laws instituted by the Creator, "I would say that the latter is greatly preferable as it implies a far grander view of the divine power and dignity than the other." Since nothing in the inorganic world "may not be accounted for by the agency of the ordinary forces of nature," why not consider "the possibility of plants and animals having likewise been produced in a natural way?" In other words, he sought to supplement natural theology by verifiable natural law(s). William Whewell, one of the Bridgewater authors, had argued similarly, but in less accessible prose. Darwin quoted Whewell on the page opposite the title page of the *Origin of Species*: "But with regard to the material world, we can at least go so far as this—we can perceive that events are brought about by insulated interpositions of Divine power, exerted in each particular case, but by the establishment of general laws."

Chambers was reviled from many quarters. Adam Sedgwick, professor of geology at Cambridge and geological mentor to Charles Darwin, lambasted the *Vestiges* in an eighty-five-page diatribe in the *Edinburgh Review*. He wrote to his friend Charles Lyell, "If the book be true, the labours of sober induction are vain; religion is a lie; human law is a mass of folly and a base injustice; morality is moonshine; our labors for the black people of Africa were works of madmen; and man and woman are only better beasts."

Sedgwick firmly identified himself with the old view of the natural world—an unchanging order ruled by a Creator who mandated a stable organization and social manners. He wrote, "The world cannot bear to be turned upside down; and we are ready to wage an internecine war with any violation of our modest and social manners. . . . It is our maxim that things must keep their proper places if they are to work together for good."

THE FLOOD

Like it or not, the world *was* in the process of being turned upside down. Within a few years Darwin would comprehensively scuttle the restricted deist notion of God as being nothing more than a clever designer. The Bible was having to be reinterpreted. Copernicus and Galileo had moved the Earth from the center of all things to that of a planet circulating around the Sun. Where was heaven in this new cosmology? And what about the age-old stories of God coming in judgment through natural disasters, like the flood?

Thomas Burnet (1635–1715) wrote *The Sacred Theory of the Earth* between 1684 and 1690. It appeared alongside two other great works published at the same time: Ray's *Wisdom* and Isaac Newton's *Principia Mathematica* (1687). Burnet's *Sacred Theory* was the most idiosyncratic of the three. (It has been said that Burnet would have become archbishop of Canterbury if he had not written it.) Burnet's *Sacred Theory* was the most influential in its time, but has been the least enduring of the three.

The *Sacred Theory* was an attempt to explain the facts of geology—why the surface of the earth is so uneven in terms of mountains, seas, and landmasses, why islands exist. Burnet began with the almost unquestioned assumption of his time that God had created the world as a series of concentric layers, with a crust lying over water, and "no Rocks or Mountains, no hollow caves, nor gaping Channels, but even and uniform all Over." Rivers ran from the poles to the tropics, where they dissipated. This primitive order disappeared in the devastation of the biblical flood. For Burnet, the Earth of his time was the shattered ruin of a "perfect" pre-flood creation; the oceans were gaping holes and the mountains upturned fragments of the old Edenic crust. He was particularly offended by mountains: "If you look upon a heap of them together they are the greatest examples of confusion that we know in Nature; no tempest or earthquake puts things into more disorder." Burnet's book greatly impressed Newton, who wrote to Burnet to congratulate him. The book was reprinted repeatedly throughout the 1700s and was regarded as a significant geological text.

The flood is, of course, tied in the Bible to the story of Noah, who took into an ark built at God's command "a male and a female of all beasts, clean and unclean, of birds, and of everything that creeps on the ground" to save them from drowning in a great deluge. The interpretation of this story has exercised biblical exegetes over many centuries: Could every species fit into the ark? Where was the fodder stored? What happened to the dung? How could carnivores and herbivores coexist? A new level of questioning arose in the seventeenth century, however, with increasing knowledge of the New World. The Americas were variously interpreted as a land that had wholly escaped from the flood, a sodden continent only recently risen above the waters, or the New Atlantis that Plato described in his dialogue *Timaeus*.

Even British jurists tried their hand at explanations. Lord Chief Justice Matthew Hale (1609–1676) proposed that American animal life had appeared through a kind of migration and subsequent degeneration—or, if not, perhaps that American species are worldwide and they would be found elsewhere when Africa and Asia are more thoroughly explored. He interpreted native Americans as descendants of the lost tribes of Israel, naked innocents who had escaped the fate of Noah's wicked generation, or perhaps degenerate savages who had wandered away after the ark grounded. In France, Isaac de La Peyrère (1594–1676) argued that the flood was a local event and that Adam was not the first-created man, but merely the first Jew, with the natives of the Americas representing other species. Hale rejected this as "immoral and irreligious" and insisted that all humans were descended from Adam, and that the same argument could apply to the animal kingdom: all animals were descended from those that came out of the Ark.

A century after Hale and La Peyrère, Linnaeus listed fourteen thousand species, fifty-six hundred of them animals. How could so many creatures have been passengers on the ark? Linnaeus regarded himself as "the publisher and interpreter of the wisdom of God" and opted for a revisionist version of the Bible. He suggested that all living beings, including humankind, originated on a high mountain at the time the primeval waters were beginning to

FIGURE 3.1. Noah's ark from the *Nuremburg Chronicle* (1543). Reconstructions of the ark from biblical data became increasingly strained in the sixteenth century as travelers began to bring to Europe a wealth of unusual creatures from newly discovered regions.

recede. Extrapolating backward, as it were, he concluded that in the beginning only one small island had been raised above the surface of a worldwide sea, which must have been the site of Paradise and the first human home.

As the mountain grew higher and higher, he proposed that it could have presented a wide range of ecological conditions, arranged as belts from tropical to polar zones. He envisaged organisms moving in turn to the latitudes where they were to remain for the rest of time. He did not contest the biblical account of creation, but questioned the ark episode; it played no part in his understanding. He asked, "Is it credible that the Deity should have replenished the whole earth with animals to destroy them all in a little time by a flood, except a pair of each species preserved in the Ark?" He believed it was a combination of sheer chance, the speed of migration, and the helping hand of God that determined exactly where each species ended up.

Under Linnaeus's scheme, the whole system was delicately

balanced, with plants and animals perfectly constructed for the environment where they lived. Linnaeus saw the natural world as a static picture, with every tiny detail preordained by the Creator God. Unfortunately for his argument, this thinking opened the door for more radical accounts. If every species was so well suited to its environment, it was difficult to imagine their members traveling across vast tracts of land from Mount Ararat. Perfect adaptation of organisms seemed incompatible with migration from a single source.

A German zoologist, Eberhardt Zimmermann (1743–1815), was ruthlessly critical of the Linnaean interpretation. He pointed out that the first pair of lions on Linnaeus's mountain would soon eat the first pair of sheep, then the goats, cows, llamas, buffaloes, zebras, and so on, in quick succession. Finally the lions would die from hunger. Zimmermann wrote that it would be far better that every animal should be created in the area where it was destined to live, under the same climate that it now enjoys, with the same food rations already in abundant supply.

Once the improbability of every animal coming together in the ark or having a common source on Mount Ararat was recognized, it became more plausible to probe further into the origin and distribution of organisms. The French Comte de Buffon (1707–1788) pointed out that there were regional associations of animals, inhabiting different areas of the globe; the Swiss botanist Augustin de Candolle (1778–1841) defined botanical "regions" where endemic (that is, native) species were found. Fauna and flora of different areas started to be written.

The idea of multiple centers of creation began to be canvassed. Increasing knowledge of the distribution of animals and plants was a major factor in leading to the recognition that living things required study in their own right—a major step to the study of the real world, as opposed to castles in the mind like Burnet's scheme. Wonderment at nature and a passion to collect flora and fauna started the final revolution that we now call ecology.

BIOGEOGRAPHY: DISTRIBUTION OF ANIMALS AND PLANTS

In the mid-eighteenth century, Carl Linnaeus set a trend for biological realism with his collections, derived from systematic soliciting of specimens from a wide range of localities. He sent out so-called "apostles" to many parts of the globe. The most famous was Daniel Solander (1733–1782), whom Linnaeus intended to be his successor. Solander, however, having gone to England to publicize Linnaeus's ideas, stayed there. Eight years after his arrival (in 1768), he was employed to accompany Joseph Banks (1743–1820) on the first voyage of the *Endeavour* under Captain James Cook. Banks bestrode British science for half a century. He was President of the Royal Society for over forty years, unofficial director and reinvigorator of the Royal Botanic Gardens at Kew, patron of many scientists and scientific expeditions, and energetic advocate of transplanting economically important species.

Specimens that Banks collected on the *Endeavour* voyage led to the recognition of 110 new genera and around 1,300 new species, but his main legacy is as an administrator rather than a practicing biologist. A younger German contemporary, Alexander von Humboldt (1769–1859) had a much greater influence on science *sensu stricto*. Humboldt traveled extensively in Central and northern South America, fired by the enthusiasm of his University of Göttingen friend, Georg Forster (1754–1794), who had sailed as a naturalist on James Cook's second voyage (replacing Banks, whose demand for accommodation was so excessive that he was refused a place on the ship).

Besides the accounts of his travels, Humboldt made observations that have formed the basis for quantitative ecology, physical geography, and meteorology. On his way across the Atlantic, he spent some time on the Canary Island of Tenerife where he described vertical zonation in the vegetation of the central mountains, the sort of analysis that today forms the preliminary stage of any ecological study. He invented isobars and isotherms as an aid to showing the limits of particular species and natural assemblages, and developed botanical arithmetic, the ratio of species in one group of plants to that in another. This ratio can show the predominant forms present in a region and the general relationships between different groups. The technique was developed further by Augustin de Candolle (1778–1841); used by Humboldt's friend, Christian von Buch (1774–1853) in his essay on the Canary Islands, which for the first time set out evidence for geographical speciation; by Charles Darwin in comparing the flora of different archipelagos; and by Joseph Hooker (1817–1911) in comparing continental and insular populations.

All of this led to a new science: biogeography. Its founding is generally attributed to the French scholar the Comte de Buffon. He was mainly a laboratory scientist, but his study of regional associations of animals showed the way forward. The major stimulus and development for biogeography undoubtedly came from Humboldt and the pioneering scientific voyages of the eighteenth and nineteenth centuries.

The early voyages were concerned almost exclusively with exploration and mapping. From the biological point of view, the first significant one was James Cook's first voyage in 1768–1771, during Buffon's lifetime; Cook took with him Joseph Banks and Linnaeus's disciple, Daniel Solander. The expedition tends to be remembered for Cook's discoveries in Australia and for Banks's collections, but its overt purpose was to establish on Tahiti a station to observe a transit of Venus across the Sun, allowing the distance of the Earth from the Sun to be calculated and hence the calibration of other astronomical data critical for navigation. It was followed by a French expedition in 1800–1804 under the command of Nicolas Baudin (1754–1803), which provided a mass of material for Lamarck's studies in Paris.

More important because of the generalizations that resulted from it, was the voyage of HMS *Beagle* under Robert Fitzroy in 1831–1836, with Charles Darwin as Fitzroy's "gentleman-companion."[1] Like

Ray before him, Darwin was entranced by the scenes he encountered: "The sense of sublimity which the great deserts of Patagonia and the forest-clad mountains of Tierra del Fuego excited in me, has left an indelible impression on my mind."

Darwin was primarily a naturalist; he was not a theoretician nor a natural rebel. He marveled at the lushness of the tropics that he first encountered in Brazil. He was impressed by the biota of South America and how it was apparently adapted to local conditions while differing from that on other continents, and how one form replaced another along the length of South America as conditions changed. He was puzzled about the numbers of unique species on oceanic islands. It was only after his return to Britain that he realized the significance of the variety of birds on the Galapagos Islands, but he noted at the time, "I never dreamed that islands about fifty or sixty miles apart and most of them in sight of each other, formed of precisely the same rocks, placed under a quite similar climate, rising to nearly equal height, would be differently tenanted. . . ."

Darwin's voyage (he never left Britain again) was followed by that of Joseph Hooker on HMS *Erebus* (1839–1843). The young Hooker idolized Darwin; he took a copy of Darwin's *Voyage of the Beagle* with him on the *Erebus*. Later in life he wrote,

> The [book] impressed me greatly, I may say despairingly, with the genius of the writer, the variety of his acquirements, the keenness of his powers of observation, and the lucidity of his descriptions. To follow in his footsteps, at however great a distance, seemed to be a hopeless aspiration; nevertheless they quickened my enthusiasm in the desire to travel and observe.

ISLANDS AND THEIR SPECIES

Good scientist that he was, Hooker was influenced by his observations. His notes on the voyage quickly expanded from the minutiae of collecting to questions of geographical distribution. Madeira, his first island, "strongly reminded me of some of the islands on the West of Argyllshire [off Scotland].... The ravines are quite like Scotch ones, but more sparingly wooded." In a letter to his father written during the journey through the South Atlantic, he clearly thought that the island biotas he would find would be determined by temperature.

By the time the *Erebus* reached Kerguelen Island in the southern Indian Ocean, midway between Africa, Australia, and the Antarctic, Hooker had begun to ask deeper questions about the relation between the flora of islands and continents. He concluded that the most marked influence on the Kerguelen flora was "Fuegian," allied to the flora of Tierra del Fuego at the tip of South America. As he journeyed on, he came to see the Fuegian flora as "the great botanical centre of the Antarctic Ocean"; all the islands south of New Zealand, the Falkland Islands, South Georgia, Tristan da Cunha, and Kerguelen "seemed to have borrowed plants" from there. He found this astonishing; Kerguelen, for example, was five thousand miles from Tierra del Fuego. Not only that, Fuegia possessed a great number of English plants. His mind was drawn to "that interesting subject—the diffusion of species over the surface of the world."

This led to a long continued debate with Darwin: Were oceanic islands vegetated as relics of now-submerged continents, or did the flora arrive by migration, as Darwin believed? This question, of course, is the heart of biogeography; it is the same problem Zimmermann faced in objecting to Linnaeus's idea of a spread from Ararat. We now know that both mechanisms are operating. Some of the best data on this point come from the colonization of recently erupted volcanoes.

Colonization is easier and turnover more rapid when empty habitats are available—this happens when a volcano forms a new island or sweeps an existing area of all its life. Some wide-ranging species quickly establish themselves whenever a possibility occurs; they have been described as supertramps—good colonizers but usually poor competitors, a distinction that recalls that between *r*-

and *K*-strategists. The island of Krakatau in Indonesia exploded in 1883. In the first phase of re-establishment of biological life, most of the colonizers away from the shore (an average of two species a year) were windborne, although nearly as many (1.64 species per year) colonized the shoreline. In the early years, only an average of one animal species every seven years penetrated into the interior. During the next phase (1897–1919), windborne pioneers lost ground and animal penetration of the interior increased tenfold (to an average of 1.32 species per year) as fruit-bearing plants grew up and provided food for immigrants.

Then in 1930 further volcanic activity produced a new island, Anak Krakatau ("Child of Krakatau"), which has experienced sporadic eruptions ever since. A visitor six months after its appearance wrote,

> Its virgin shore [is] composed entirely of dark grey ash, black cinders and white pumice stone. No plants would grow here until weathering and bacteria had had time to create soil in a year or two, but seeds, along with debris like banana stems and other vegetable matter, were awaiting their time to establish themselves. . . . The only abundant insects were scavengers which could feed on whatever plant life the sea brought them—a springtail, a beetle, a species of ants, a tiny leaf-mining moth and a mosquito.

The first birds seen were a Pacific reef-egret and a beach thick-knee, together with migrants like the common sandpiper, grey wagtail, Pacific golden plover, Mongolian plover, whimbrel, and great knot. In 1985, seventy-two species of small flying creatures were collected in ten days in plastic containers filled with seawater in places where the lava had flowed—a constant rain of arthropods. A great increase in bird species occurred in the same year, coinciding with the first fruiting of the island's figs and the consequent opening of a new "habitat window." As these immigrants established themselves, some of the earlier-arrived ground-nesting species disappeared. Then raptors came to feed on the fruit-eating species. About 150 plant species now inhabit the island. The same process has been occurring on Surtsey, an island produced by a volcanic eruption off the southwest corner of Iceland in 1963, although here the land surface has been eroding so quickly that the succession has not been as marked as for Krakatau (see Figure 3.2).

One can also study colonization experimentally. Ed Wilson and Dan Simberloff painstakingly undertook such a study, fumigating four small mangrove islands off the coast of Florida so as to kill all the resident animals and then monitoring the islands' recolonization over a period of years. Intriguingly, in situations that are intuitively stable—mature forest or rocky seashore—a similar pattern of colonization to that following volcanic eruptions occurs. University of Washington ecologist Bob Paine has followed the sequence of events after an old tree has fallen or a patch of shore has been artificially cleared of its animals and plants. He found that the dynamics of the reestablishment of the fauna and flora are the same as in the more spectacular circumstances of Krakatau and Surtsey, albeit in a less obvious manner.

A major advance in the understanding of island biotas has been "the theory of island biogeography" that Robert MacArthur and Ed Wilson put forward in 1963 and expanded into a book in 1967. The core of their thesis was that a balance exists between immigration to an island determined by its distance from the mainland and extinction of local populations, which will vary with the island area. In other words, the number of species on an island will be the difference between those continually reaching it and those that are being lost (see Figure 3.3). Their insight was that the species composition is a present dynamic, not a hangover from something that already happened. Species are continually going extinct locally; species are continually appearing and establishing themselves.

What MacArthur and Wilson expressed as a quantitative theory

FIGURE 3.2. a) Surtsey: An island in the Westmann group off the south coast of Icleand formed in 1963 as the result of an undersea volcanic explosion (Photo: Sturla Fridriksson). b) Pioneer shore community of lyme grass and sea sandwort on sandy lava on Surtsey. These species may have come before from a neighboring island or the mainland of Iceland (Photo: Borgthor Magnusson).

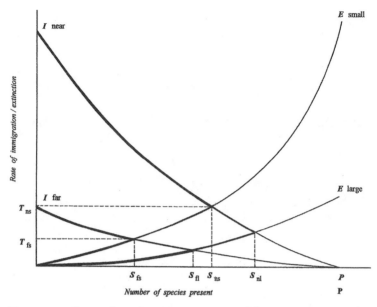

FIGURE 3.3. The number of species on an island (S) is the result of a turn-over (T) between rates of immigration (I) and extinction (E). Immigration is affected by the proximity of an island to its source of colonizers (n = near, f = far) and the size of the island (s = small, l = large) (based on Robert H. MacArthur and E. O. Wilson, *Theory of Island Biogeography* [Princeton, NJ: Princeton University Press, 1967]).

is really what naturalists have intuitively known for a long while. English naturalist Alfred Russel Wallace, codiscoverer with Darwin of the idea of natural selection, described the same idea in his *Island Life* (1881):

> The distribution of the various species and groups of living things over the earth's surface and their aggregation in definite assemblages in certain areas is the direct result and outcome of... firstly the constant tendency of all organisms to increase in numbers and to occupy a wider area, and their various powers of dispersion and migration through which, when unchecked, they are enabled to spread widely over the globe; and secondly, those laws

of evolution and extinction which determine the manner in which groups of organisms arise and grow, reach their maximum, and then dwindle away....

British botanist H. C. Watson put numbers on the principle. He pointed out that a single square mile of the English county of Surrey held half the plant species found in the 650 square miles (1700 km^2) of the county as a whole. Henry Gleason generalized this in a 1922 paper "On the Relation between Species and Area." Then, in the 1950s, Philip Darlington applied the same principle to islands. He noted that in a range of islands, when the area was increased by tenfold, the number of species merely doubled. He also found that the species present in any one place are not fixed; some animals or plants disappear while others appear. All of these studies point to the same scenario: a coming and going of species, and a tendency for them to aggregate in certain places.

It was left to MacArthur and Wilson to give formal expression to these ideas in the 1960s. They suggested that recurrent colonizations and extinctions create an equilibrium, in which the number of species remains relatively constant although the species concerned vary over time (see Figure 3.3). MacArthur and Wilson also used data from the recolonization of Krakatau to support their thesis.

Data about extinction and recolonization are more accurate for birds than for any other group, because it is usually easy to record new breeders or species failing to breed. The reason for failure may be because numbers are declining generally or because of local factors like competition, change in habitat, or even pure chance. One complication is that different species have different mobility, whatever their potential powers of flight. For example, woodpeckers rarely cross water, although obviously they may be involuntary colonizers through accidents of weather or other rare events. The ancestors of the finches on the Galapagos or the honey-creepers on Hawaii cannot be said to have "intended" to settle where they ended up, and their successful establishment and breeding represent extremely unlikely events. It is impossible to know how many of their relatives perished at sea, although some indication is given by twitchers' delights—"vagrant" birds from distant parts of the world that appear on distant shores to the excitement of fanatical bird-watchers ("twitchers")—albeit mostly to die a lonely death.

MacArthur and Wilson described the balance between immigration and extinction as a state of equilibrium. In fact, it is really nothing more than a logical necessity. The number of species on an island can only be increased by two processes—immigration, which in turn depends on the distance of the island from the source of potential colonizers and the availability of ecological space for them—and be decreased only by those that fail to survive, that is, by extinction. Certainly, the theory needs supplementing with ecological information. For birds, the idea of optimal foraging improves the theory's predictions, since organisms are more likely to stay longer in an area if the distance to a neighboring island is great.

GENETIC VARIATION: THE FOUNDER EFFECT

In the 1970s and 1980s, a major change in population biology took place. This change has radically affected the way we think of the way in which species that are native to an area—called endemic species—originate. Biologists traditionally thought of individual animals or plants as genetically rather uniform: all members of a species had roughly the same genes and passed these on to their offspring without much change. Clearly, some inherited variation occurs; we see this in variations such as bridled murres, black rabbits, pin *versus* thrum primroses, and mammals with different blood types. But in the past, biologists believed that the proportion of variable gene loci—that is, genes that have slightly different forms in a population (technically called "alleles")—was thought to be very small. Indeed a simple calculation apparently showed that too much genetic variation could not be tolerated: it produced a *genetic load* that reduced fitness and crippled the reproductive potential of the population.

Today we know that this assumption of little variation—near genetic homogeneity—is too simplistic. It has had to be revisited because of experimental results. In 1966 Harry Harris (working in London on human material), and Jack Hubby and Richard Lewontin in Chicago (working on the fruit fly *Drosophila pseudoobscura*) demonstrated that considerable genetic diversity existed. Such diversity can be most easily measured by heterozygosity, the proportion of genes in an individual where different alleles are inherited from its parents. Many studies have shown that heterozygosities of 10 percent or more are commonly found in a wide range of organisms (see Table 3.1).

TABLE 3.1 **Percentage of Heterozygous Genes per Individual (i.e., Different Forms [Alleles] Inherited from Each Parent) as Measured by Protein Electrophoresis**

Drosophila		15.0
Other insects		15.1
Land snails		15.0
Marine snails		8.3
INVERTEBRATES		14.6
Fish		7.8
Amphibians		8.2
Reptiles		4.7
Birds		5.4
Mammals:	Rodents	5.4
	Large Mammals	3.7
VERTEBRATES		5.0
PLANTS		17.0

Different classes of genes have different levels of heterozygosity, but all show high levels.

Two consequences of this high heterozygosity are extremely important in understanding how an island species may differentiate its originating population. First, a small group of individuals drawn from a large population will almost certainly differ from its parental group in the frequency of alleles at a large number of genes. Second, some alleles will be absent or relatively over-represented in the descendant group. This small-group effect will be particularly important if only a small band of organisms manages to colonize an empty island or habitat. This daughter group will be immediately different from its source population, which is, of course, its parents. These effects only became evident after the time that MacArthur and Wilson were developing their ideas and probably explains why they erred in their discussion of the origin of island species.

To be fair to MacArthur and Wilson, they were not particularly interested in the processes of speciation. But what they did say about evolution of species takes us to the next stage of the debate about speciation. The animals or plants that colonize an empty habitat constitute the founders of a new population. Their characteristics lead us to consider the *founder effect* or principle. MacArthur and Wilson touched on this question in a chapter on evolutionary changes in their famous book. They begin by saying that since "we believe that evolution through natural selection has produced the biotic differences which characterize islands, it is appropriate for us to study how natural selection works on islands." They continue:

> We can think of the evolution of the new population as passing through three overlapping phases. First the population is liable to respond to the effects of its initial small size. This change, if it occurs at all, will take place quickly, perhaps only in a few generations. The second phase, which can begin immediately and must continue indefinitely, is an adjustment to the novel features of the invaded environment. The third phase, an occa-

sional outgrowth of the first two, consists of speciation, secondary emigration and radiation.

MacArthur and Wilson explicitly equated their first phase with the founder effect, a concept put forward in 1942 by the Harvard taxonomist and evolutionist Ernst Mayr in one of the defining works of the neo-Darwinian synthesis (see p. 107), but described more fully in a 1954 essay "Change of genetic environment and evolution." For their part, MacArthur and Wilson regarded the founder effect as "an omnipresent possibility but one easily reduced to insignificance by small increases in propagule size, immigration rate, or selection pressure.... The founder principle is actually no more than the observation that a [founding] propagule should contain fewer genes [alleles] than the entire mother population."

This is wrong, however. It would have been a reasonable assumption before the discovery of high levels of heterozygosity, but the post-1966 revolution showing the enormous genetic variability in any group of organisms means that the founder effect will almost certainly change allele frequencies as well as reduce variability to some extent. The geneticist Sewall Wright, one of the founders of modern population genetics theory, makes this clear. He pointed out that Mayr used founder theory to explain only the loss of genetic variability, not its distribution. He referred to Mayr's emphasis on the founder effect as leading to gene (or allele) loss and a reduction in variability, but for Wright the "most significant" genetic feature in the founding of a species was

> the wide random variability of gene frequencies (not fixation or loss) expected to occur simultaneously in tens of thousands of loci at which the leading alleles are nearly neutral, leading to unique combinations of gene frequencies in each of innumerable different local populations.... The effects attributed to the "founder effect" by

frequencies of allelomorphs at one locus in population

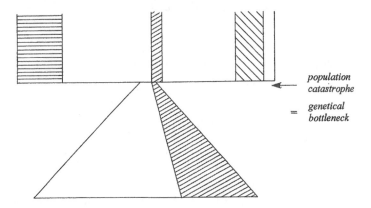

population catastrophe

= *genetical bottleneck*

FIGURE 3.4. Effects on the frequencies and numbers of alleles at a single gene locus as the result of a bottleneck in numbers of a population. This effect is especially marked when a small number of individuals found a new population. Similar loss of alleles and changes in frequency occur at every locus in the genome, resulting in the founded population being genetically different from the ancestral one.

Mayr [and also by MacArthur and Wilson] are the most obvious but the least important of the three.

In other words, when a small population colonizes an empty habitat like an island, that population will almost inevitably be different from its source population. The effect is that every colonizing event is a new experiment, exposing to the environment a set of reaction systems determined by the alleles carried by the founding group. It is these systems that represent the second generation's ability to respond (or not to respond, given that most colonizations result in rapid extinction because of the failure of the animals or plants to cope in their new situation) to natural selection. This response is phase 2 in the MacArthur and Wilson scheme; it will necessarily be limited by the chance collection of alleles present in the founders.

This "instant" result of a founding event is obvious once the effect of taking a small number of individuals from a genetically heterogeneous source is realized. Some theoreticians still doubt the importance of founder effects in speciation, but a growing number of studies show their effect in establishing local genetic heterogeneity and hence the possibility of further differentiation. Indeed the founder effect has been described as perhaps the most novel and influential contribution of the twentieth century to ideas about how evolution occurs on islands. There seems no reason to doubt that it is the main determinant of the distinct island races.

It is probably helpful to distinguish the founder principle, which occurs through the chance collection of genes carried by founders, from founder selection, which occurs following isolation; the founder effect is the overall result of both processes. Separating the two is usually impossible because we rarely know the frequencies of genes carried by the founding group. However, reconstructing the genetic constitution of the founders in some human isolates has been possible.

One example comes from the human population of Tristan da Cunha, a volcanic island in the South Atlantic Ocean. The island has the highest frequency in the world of a recessively inherited progressive blindness, retinitis pigmentosa. Fifteen individuals effectively founded the Tristan population. We know the pedigree of them all from the time the community was established in 1817 (when the military garrison intended to prevent the escape of Napoleon from the sister island of St. Helena was withdrawn) to the time the population was evacuated in 1961 following the eruption of the island volcano. One of the original founders of the human population must have been carrying an allele of the gene affecting retinitis pigmentosa. In technical language he or she must have been heterozygous for the *retinitis pigmentosa* mutation. Since we all get a set of chromosome from each of our parents, the frequency of that mutation was at least one in thirty—about 3 percent. The population increased seventeen-fold during its period of isolation, but the frequency of the allele remained constant. There were seventeen copies of the allele in 1961—four homozygotes with two copies each of the gene and nine heterozygotes, each with a single copy.

In another case, founder selection seems to have been acting as well on a condition called *porphyria variegata*, an inherited defect of porphyrin, part of the hemoglobin molecule. The porphyria allele causes extreme sensitivity to sunburn, so its occurrence is easily recognized. In South Africa, it is carried by about eight thousand individuals, three in every thousand of the descendants of the original white population, but is very rare outside the country. All the present-day sufferers in South Africa are in thirty-two family groups, all of whom can all trace their ancestry back to one of the original forty pairs of white settlers who arrived from Holland in 1688. One million members of the current population have the same surnames as the original forty—implying an increase of 12,500 times in three centuries. However, the porphyria allele is present in only two-thirds of these. Presumably its deleterious effects (particularly in the sunny climate of South Africa) have led to selection against porphyria carriers whether through leaving South Africa, dying from the condition, or failing to have children.

The Case of the Spittle-Bug

One of the best examples of selection taking place after the original founder event is in a series of experiments involving the froghopper or spittle-bug, *Philaenus spumarius*, on small islands in the Baltic Sea off the coast of Finland. The spotting pattern on the wingcases (*elytra*) of these bugs distinguishes a range of forms (or *polymorphisms*), each determined by an allele at a single gene locus. On the mainland of Finland, one form of spotting pattern on the wings is more common toward the north, while another form has a higher frequency in more humid areas, implying that the insect wings are adapting to different conditions.

The island races of the spittle-bug tend to reflect the variability found in mainland populations. Although several of the wing casing (elytral) forms are missing on the outer islands, the same forms also disappear at the northern edge of species range on the mainland, despite the fact that the bugs there are in breeding contact with a large southern population. In other words, the genetic structure of the island races seems to be based on their adaptation to available plant life, not merely the result of more of the bugs arriving or leaving the areas.

Olli Halkka, a Finnish biologist, and his colleagues found that some forms of the bug seemed to be particularly hardy and survived better than others when they were introduced to islands where they did not occur. But the survival advantage changed as the vegetation changed. The original colonizers were replaced by another form that had a preference for feeding on newly dominant plants, like purple loose-strife (*Lythrum salicaria*) and sea mayweed (*Triplospermum maritimum*).

Direct proof of selection on the islands came from an experiment in which approximately eight thousand individuals (three-quarters of the populations) were swapped between two islands with genetically different populations. After three generations, both island populations had reverted to the pre-transfer allele frequencies, although they deviated considerably from another thirty-five island populations in the area. In other words, the genetic makeup of the island populations was not random (i.e., determined by genes introduced by the founding animals) and must therefore be regulated by natural selection. These Finnish island spittle-bugs are one of the few cases where separating the genetic founder effect from subsequent selection has been possible.

At this point, we return to Joseph Hooker. Islands brought Hooker and Darwin together. Darwin wrote to Hooker soon after the latter disembarked from his four-year round-the-world voyage on the *Erebus*:

> I had hoped before this time to have had the pleasure
> of seeing you & congratulating you on your safe return

from your long & glorious voyage. I am anxious to know what you intend doing with all your materials—I had so much pleasure in reading parts of some of your letters [sent to Charles Lyell, Senior, father of the geologist], that I shall be very sorry if I, as one of the Public, have no opportunity of reading a good deal more.... Henslow [Darwin's botanical teacher] (as he informed me a few days since by letter) has sent to you my small collection of plants—You cannot think how much pleased I am, as I feared they wd have been all lost. I paid particular attention to the Alpine flowers of Tierra Del. & I am sure I got every plant which was in flower in Patagonia at the seasons when we were there.— I have long thought that some general sketch of the Flora of that point of land, stretching so far into the southern seas, would be very curious.

Hooker responded quickly:

I am exceedingly glad to think you attach so much importance to the comparison of the Arctic plants with the Antarctic as it was my aim throughout to establish an Analogy between the two hemispheres, & to draw up tables upon several plans, shewing for instance the proportion of plants in each of the predominant Nat[ural] Ord[er]s. common to both.... In my Antarctic flora I intend (following my fathers advice) to include Ld Aucklands & Campbells Islds as they contain the most southern plants of those longitudes, & as they have all the nameless peculiarities of plants of high latitudes, quite as much so as those of Fuegia (however luxuriant the vegetation may be compared with analogous Northern latitudes).... The Vegetation of Kerguelens Land is entirely that of Southernmost America, almost all its plants

being common to the two, few in proportion common
to it & Ld Aucklands & none peculiar to the two latter
(perhaps one is). The Falkland Isld. flora seems to combine the Patagonian with the Fuegian, I think of including it with the latter.

Darwin and Hooker remained friends and mutual critics. Hooker
was the first person told about Darwin's evolutionary ideas. He
spoke at the infamous debate in Oxford in 1860 between T. H. Huxley and the bishop of Oxford. Hooker also set out definitive principles about island biotas during the 1866 meeting of the British
Association for the Advancement of Science. His lecture was his-
torically significant in the scientific support Hooker gave for Darwin's *Origin of Species*, but has continuing relevance as a penetrating
analysis of an important evolutionary situation. The lecture was the
first systematic statement of the importance of islands for evolutionary studies. Hooker's identification of the main characteristics
of island biotas still stands:

▸ They contain a high proportion of forms found nowhere else
(endemics), although these endemics are usually similar to
those found on the nearest continental mass.

▸ They are impoverished in comparison with comparable continental areas, that is, there are fewer species on islands than on
mainlands.

▸ Dispersal must play a part in the colonization and establishment of islands, unless the island has been cut off from
a neighboring area and therefore carries a relict of a former
continuous fauna and flora.

▸ The relative proportions of different taxonomic groups on
islands tends to be different from nonisland biotas—that is,
there is taxonomic "dysharmony."

Peter Grant, who through his careful and long-continued studies
on "Darwin's finches" on the Galapagos has probably done more
than anyone else to elucidate evolution on islands, has written, "An

outstanding feature of islands is their strangeness; many of them are downright weird. Naturalists of the last three centuries brought back to civilization accounts of strange and unimagined creatures found only on remote islands. Dodos. Sphenodon. The Komodo dragon. Daisies as tall as trees. What is it about islands that promotes such strangeness?" Hooker did not know the answer to this, but he was a pioneer in beginning the search, and his work inspired—and continues to inspire—his successors. Through his start, we can now start to answer reasonably Grant's question about the curiosities of so many of the plants and animals on islands.

CHAPTER 4
Stewardship and Ecological Services

The insufferable arrogance of human beings is to
think that Nature was made solely for their benefit,
as if it was conceivable that the sun had been set
afire merely to ripen men's apples and head
their cabbages. —Cyrano de Bergerac

WE TEND TO assume that the natural world is available for us
to use as we please. In past centuries more than today, we might
have regarded our use under some sort of divine authority, but we
always regarded nature as subservient to us. There was an implicit
hierarchy: God-humans-nature. This was fine in theory, but dominion over nature has problems. A society that assumes its right to
dominion over nature is faced with a paradox: fullest exploitation
of the natural world involves its eventual destruction; the destruction of nature involves the destruction of society itself.

This is the knife edge on which the management of all biologically based resource systems rests. If successful management per-
mits a steady harvesting over the longest possible period of time,
maintaining a level of production without deterioration in the system itself may be possible. This harvest would be greatly increased
and a much higher rate of production achieved if the ideal of sustained yield is abandoned, but this could only be in the short term.
It is like money in the bank: you can live off the interest or you can
take from capital. If the exploitive process is arrested before the system is completely ruined (i.e., all the capital is gone), a period of

rest may permit a return to a fully productive condition, but this recovery must by no means be taken for granted. Most, probably all, ecosystems have a point of no return beyond which restoration is no longer possible.

The ancient Hebrews achieved a balance with a concept of mankind's responsibility to God for the management of the Earth, a concept that was carried over into Christianity, becoming part of the Western heritage. This concept of stewardship of the Earth is also embedded in Islam. In the Qur'an, before creating Adam, God says to the angels (who were initially rather hostile to the suggestion), "I am going to place a *khalifah* on earth." *Khalifah* can be translated as "substitute," but "vice-gerent" or "deputy" is probably a better translation. A steward is one who cares for property on behalf of another; a good steward is one who strives to avoid any damage or deterioration in that entrusted to him or her.

The idea that we are responsible to God for the use of the Earth was an integral part of Judeo-Christian thought for many centuries (probably less so in Islam). It was implicit in the Benedictine rule and was part of its success in establishing sway over large tracts of land in Europe before the Reformation. Someone we have already met (p. 76), Sir Matthew Hale, a chief justice of England, expressed the idea explicitly in his book *The Primitive Origination of Mankind* published in 1677:

> In relation to this inferior World of Brutes and Vegetables, the End of Man's Creation was, that he should be the VICE-ROY of the great God of Heaven and Earth in this inferior World; his Steward, Villicus, Bayliff, or Farmer of this goodly Farm of the lower World. . . . If we observe the special and peculiar accommodation and adaptation of Man, to the regiment and ordering of this lower World, we shall have reason, even without Revelation, to conclude that this was one End of the Creation of Man, namely, to be the Vice-gerent of Almighty God,

in the subordinate Regiment especially of the Animal
and Vegetable Provinces.

In Hale's view we could not escape from our responsibility to God
for the proper management of the Earth, with the task to control the
wilder animals and to protect the weaker, to preserve and improve
useful plants and to eliminate weeds, and interestingly, to maintain
the beauty as well as the productivity of the Earth. It was an important idea, but one implicitly challenged and effectively overturned
through a rationalization proposed by John Locke (1632–1704), to
justify accumulation of land additional to the needs of the owner.
Locke developed a theory of property based on the argument that
a person's labor belongs to him or her, to do with it whatever he
or she wants. Since an individual owns his or her own labor, society is not involved; if the right to unlimited ownership of property
depends on personal labor, property rights carry no social obliga-
tion. This meant that there need be no objection to the right to own
unlimited property. It was an argument that degraded stewardship
responsibilities and effectively opened the way to the Industrial
Revolution. People changed from being responsible stewards and
God's image-bearers to being "human resources."

The Industrial Revolution spawned social upheavals (and
unrest), and rafts of legislation to regulate the exploitation of work-
ers. It also raised concerns for the countryside. In Britain, the General Inclosure Act of 1845 is sometimes described as the first piece
of conservation legislation. At a time when urbanization was pro-
ceeding rapidly and fortunes could be made by enclosing and then
selling urban plots for building, the Inclosure Act acknowledged
that enclosure was the concern of all the local inhabitants, not
merely of the lord of the manor and a privileged group of com-
moners; the health, comfort, convenience, exercise, and recreation
of all the local inhabitants should be taken into account before any
enclosure was sanctioned. It marked an awareness of what became
to be known as the "environment."

It would be wrong to regard environmentalism as a simple re-action to the excesses of industrialization and technology. Environmental crises have occurred in many cultures and throughout history. Plato described deforestation in ancient Greece. The great Mayan cultures of Meso-America collapsed around the year AD 800 due to deforestation and soil erosion. Clarence Glacken has documented the interactions of human populations with their environment over many centuries to show how realization of a highly ordered created world was usually accompanied by a corresponding desire to live in accord with it.[1] Trees were cleared wherever there was human settlement. The consequent soil erosion and effects on drainage patterns were clear in early nineteenth-century Europe. In 1864 George Perkins Marsh (1801–1882), U.S. ambassador to Italy from 1861 until his death, wrote of his observations there:

> The ravages committed by man subvert the relations and destroy the balance which nature had established between her organized and inorganic creations; and she avenges herself upon the intruder by letting loose upon her defaced provinces destructive energies hitherto kept in check by organic forces destined to be his best auxiliaries, but which he has unwisely dispersed and driven from the field of action. . . . The earth is fast becoming an unfit home for its noblest inhabitant.[2]

Marsh's book was very influential. In the year his book appeared, the Yosemite Valley was ceded to the state of California as a public park, and six years later Yellowstone in Wyoming was established as a national park. During his term as president (1901–1909), Theodore Roosevelt stressed the need for scientific management of the nation's resources. He created an active forestry service under Gifford Pinchot. Science and technology came to be seen as giving mankind an ever greater control over nature; the idea of human responsibility and "stewardship" seemed decreasingly important, particularly when it seemed to involve an absent and impotent landlord.

In 1956 Stanford University historian Lynn White (1907–1987) lectured on the "Historical roots of our ecologic crisis" to the American Association for the Advancement of Science. His short and easy-to-read essay became one of the most influential and controversial theses about the environment of the twentieth century.[3] It was published in the same year as the comprehensive work of Clarence Glacken, and largely eclipsed it. White looked back behind the rationalism of Locke and the Enlightenment and the hubris of the industrial age. His area of expertise was the role of technological innovation in the Middle Ages. For White, the early Middle Ages was the defining period in the growth of Western supremacy.

There is no doubt that Christian understanding of the natural world changed during the Middle Ages. The Hebrew tradition had been that the world belonged to God, but was separate from him and not sacred. Christian doctrine had originally affirmed this, but was clouded by an emerging neoplatonism, particularly that of Origen (185–254) and Plotinus (205–270). Their world was dominated by a hierarchical pyramid of living beings with God at its apex, reminiscent of Aristotle's "Ladder of Nature" where different rungs were marked by different levels or amounts of "soul."[4] Origen saw God as creating the material world as a kind of gracious act, to stop the descent of rational spirits toward ultimate nonbeing. His thinking was dominated by the fall of humankind; for him, the fundamental reason for creation was to counter the effects of the fall. Matter was created by God and exists largely for the purpose of educating humans through trials and tribulations, before returning to a higher spiritual destiny. This meant that the world is a sort of purgatory, reminiscent in different ways to the excesses of hellfire preaching by some Protestant churches and the reincarnationalism of some Eastern religions.

Origen-type thinking permeated the early Church. By shifting attention from God's redemption of the universe to the salvation of the individual sinner, Origen introduced a distinction between creation and God's saving work. This separation was the ground for White's assertion that "in its Western form, Christianity is the most anthropocentric religion the world has seen.... Christianity, in absolute contrast to ancient paganism and Asia's religions (except, perhaps, Zoroastrianism) not only established a dualism of man and nature but also insisted that it is God's will that man exploit nature for his proper ends."

This dualism has had a profound effect on human attitudes to the Earth. Through the cultivation of land for agriculture—including the draining of wetlands and consequent changes to the water table by irrigation, through deforestation, and through the removal of prey and pest species by hunting and chemical poisons—we have had a massive impact on natural systems for our own purposes. In the early centuries of agriculture, these effects were local and limited; their influence was slight. They have increased enormously as technology has advanced.

This was White's starting point. He argued that a direct link existed between agricultural technology and social structures: the replacement of the scratch plow, which produced shallow furrows and was only usable in light soils, with a new design incorporating a plowshare and allowing much deeper furrows and productivity, had the effect of needing a team of up to eight oxen to pull it. Under the old system, fields were distributed in family units, presupposing subsistence farming; to use the newer and more efficient plow, farmers had to pool their oxen into large plow-teams. Land distribution became related to the capacity of a power machine to till the soil rather than the needs of a family. Farmers had previously been part of nature; now they were exploiters of nature.

White drew his evidence for this in illustrated calendars. In earlier times, the months were shown as passive events; after c. 830 the calendars show people coercing the world around them—plowing, harvesting, chopping trees, butchering pigs. Scholasticism nourished this duality. Alexander Neckham (1157–1217) and Bartholomew the Englishman (c. 1260) filled their writings with fantastic emblems of heraldry, with wonderful tales of plants and beasts, often based on some record of Pliny (AD 23–79) or the *Physiologus* (written in Alexandria around the second century AD); generations of preachers and storytellers used these as illustrations of morality.

This tradition of reading a moral purpose into the natural world reached its fullest expression in Thomas Aquinas (1227–1274). Through his doctrine of natural law, Aquinas extended Aristotle's belief in the rationality and purposiveness of the created order. God not only caused the universe to come into being but for Aquinas he was also the origin of the purpose that every kind of being pursues in the course of its life. Religion did not so much encroach into ecology as envelope it. Lynn White saw this producing an attitude that:

> We are superior to nature, contemptuous of it, willing to use it for our slightest whim.... We shall continue to have a worsening ecological crisis until we reject the Christian axiom that nature has no reason for existence but to serve man..... Since the roots of our trouble are so largely religious, the remedy must be essentially religious, whether we call it that or not.... What we do about our nature depends on our ideas of the man-nature relationship.

White's interpretation became widely accepted. The pioneer conservationist Max Nicholson's answer to environmental damage was brutal: "The first step must be plainly to reject and to scrub out the complacent image of Man the Conqueror of Nature, and of Man Licensed by God to conduct himself as the world's worst pest." Max Oelschlaeger, a professor of philosophy in Texas, confessed to a prejudice against religion because of the persuasiveness of White's argument—but then a repentance when he saw the flaw in White's logic: Oelschlaeger realized that the source of environmental problems was human greed, not Judeo-Christian theology.

White's own suggestion for the way forward was to turn to Francis of Assisi, who "tried to substitute the idea of the equality of all creatures, including man, for the idea of man's limitless rule of creation"—in other words, replacing anthropocentricity (interpreting everything from the human viewpoint) with biocentricity (all life regarded as equal).

What Is Biocentrism?

The earth is what we all have in common.
—WENDELL BERRY

How do we see our relation to the natural world? Anthropocentrism starts from the premise that human beings are the only beings with moral concerns, and therefore are the only proper objects of those concerns. If we move away from such a human-centered perspective, we effectively give some kind of intrinsic or noninstrumental value to things other than human beings. We can, of course, show consideration for animal welfare so as not to upset children or—more likely—because a healthy animal is more useful to us than a sick one. But if we go further than this and give status to animals because of their own being rather than their value to us, we move from anthropocentricity toward biocentricity.

An extreme expression of biocentricity is *ecosophy* or *deep ecology*, set out forcefully by the Norwegian philosopher Arne Naess (1912–2009). For Naess and the exponents of deep ecology he inspired (especially George Sessions and Bill Devall in North America and Warwick Fox in Australia), scientific ecology needs replacing by a "core democracy in the biosphere," meaning "the equal right of every form of life to live and flourish." Deep ecology involves:

▶ Rejection of the human-in-environment image in favor of "a relational, total-field image."

▶ Biospheric egalitarianism—in principle. The "in principle" qualification was inserted because "any realistic praxis necessitates some killing, exploitation, and suppression."

Naess claimed that his deep ecology did not emerge from ecological science, although he argued that the crux of deep ecology "reflects the insights of the field ecologist." He believed that ecology suggested, inspired, and fortified a set of views that had a source beyond scientific logic, facts, or induction. He set out eight differences between deep and what he called "shallow" ecology, which to him involved reductionism and disagreement about metaphysical questions (see Table 4.1).

A less extreme and less contentious approach to intrinsic value has been developed by Colorado State University philosopher and former Presbyterian pastor Holmes Rolston. Rolston accepts that we must recognize intrinsic and instrumental values but adds what he calls "systemic" value. He believes that nature itself "is a source of values, including our own. Nature is a generative process to which we want to relate ourselves and by this to find relationships to other creatures. Values include far more than a simplistic human-interest satisfaction. Value is a multifaceted idea with structures that root in natural sources."[5] In this he links to the University of Wisconsin environmentalist Aldo Leopold, who told us to "quit thinking about decent land use as solely an economic problem. Examine each question in terms of what is ethically and aesthetically right, as well as economically expedient. A thing is right when it tends to preserve the integrity, stability and beauty of the biotic community. It is wrong when it tends otherwise." Leopold called this "a land ethic."

A completely different approach to intrinsic value is that proposed by the British chemist Jim Lovelock as *Gaia*.[6] Lovelock developed his theory following a request from NASA to devise a test for detecting life on Mars. He reasoned that the atmosphere of a lifeless planet must be in equilibrium with the physical composition of that planet, and hence would consist mainly of carbon dioxide, with a small amount of nitrogen and almost no oxygen. Such a planet would have a very high surface temperature due to the blanketing (or greenhouse) effect of the carbon dioxide. Any deviation away from this equilibrium situation would indicate the presence of a disturbing influence, which could be regarded as "life." He defined life as "a member of a class of phenomena which are open or continuous systems able to decrease their internal entropy at the expense of substances or free energy taken in from the environment and

TABLE 4.1 Differences between Shallow and Deep Ecology

SHALLOW ECOLOGY	DEEP ECOLOGY
Natural diversity is valuable as a resource for us.	Natural diversity has its own (intrinsic) value.
It is nonsense to talk about value except as value for humankind.	Equating value with value for humans reveals a racial prejudice.
Plant species should be saved because of their value as genetic reserves for human agriculture and medicine.	Plant species should be saved because of their intrinsic value.
Pollution should be decreased if it threatens economic growth.	Decrease of pollution has priority over economic growth.
Third-world pollution growth threatens ecological equilibrium.	World population at the present level threatens ecosystems, the major threat being posed by the population and behavior of industrial states more than by those of any others. Human population today is excessive.
"Resource" means resource for humans.	"Resource" means resource for living beings.
People will not tolerate a broad decrease in their standard of living.	People should not tolerate a broad decrease in the quality of life but should be ready to accept a reduction in the standard of living in overdeveloped countries.
Nature is cruel and necessarily so.	Humankind is cruel but not necessarily so.

After Andrew Brennan, *Thinking about Nature* [London: Routledge, 1988], 141 subsequently rejected in a degraded form." Mars turned out to have an atmosphere precisely as expected from its geological structure, but—and this was what started Lovelock thinking—the Earth's atmosphere is radically different from expectation.

The traditional assumption of the origin of life on Earth about 4,000 million years ago is that it was driven by chemical processes involving adaptation to contemporary atmospheric conditions which were originally reducing and allowed the formation of complex molecules but then became oxidizing (see p. 10). Lovelock turned these ideas upside down; his proposal was that the atmosphere changed in response to the life developing in it. In other words, life (or the biosphere) regulates or maintains the climate and the atmospheric composition, and thus provides an optimum for itself. If this is true, the whole geobiochemical system can be regarded as a single, gigantic, self-regulating unit. William Golding (Nobel Prize–winning novelist and author of *Lord of the Flies*, and a neighbor of Lovelock's) suggested the name "Gaia" for this system, after the Earth goddess of ancient Greece.

Gaia can be treated as a scientific or as a metaphysical theory. As the former, it has been a great success whether or not it is true, because of the research it has stimulated. Lovelock has claimed five successes for the theory that the Earth is a kind of self-regulating system:

1. Prediction of the lifeless state of Mars—made in 1968, confirmed in 1977.
2. Carbon dioxide influence on climate through the biological weathering of rock—predicted in 1981, shown in 1989.
3. The constant 21 percent of oxygen in the atmosphere for the last 200 million years could be due to natural fires and phosphorus cycling, for which the biological input is important.
4. The transfer of elements necessary for life on land—predicted in 1971 to be mediated through algae, shown in 1973.
5. A link between dimethyl sulfide produced by phytoplankton in the deep oceans and cloud cover—confirmed 1987.

The most serious criticisms of the Gaia hypothesis have come from biologists. For example, Richard Dawkins wrote in his 1982 book *The Extended Phenotype*,

> Homeostatic adaptations in individual bodies evolve because individuals with improved homeostatic apparatuses pass on their genes more effectively than individuals with inferior homeostatic apparatuses. For the analogy [that the whole Earth is equivalent to a single living organism] to apply strictly, there would have to be a set of rival Gaias, presumably different planets. Biospheres which did not develop efficient homeostatic regulation of their planetary atmospheres tend to go extinct. The Universe would have to be full of dead planets whose homeostatic regulations had failed, with, dotted around, a handful of successful well-regulated planets of which Earth is one.

Lovelock responded to this criticism by devising a computer model he called Daisyworld, which showed, he believed, that the regulatory behavior he postulated for Gaia could develop simply as a property of the complex processes that link organisms to their environment. There is certainly truth in the idea that living organisms (as defined in their normal survival-driven way as opposed to the Lovelockian way) modify their environment, and this can lead to natural selection.

The scientific study and debate about Gaia is an ongoing process, but one that must be distinguished from the metaphysical speculations often linked to Gaia. The physicist Fritjof Capra has welcomed the emergence of Gaia as a sign of a universal change of attitude, so that the Earth "not just functions *like* an organism, but actually seems to be an organism. The new paradigm is ultimately spiritual." Lovelock seems equivocal about this. He has written, "For every letter I got about the science of Gaia [following his original pronouncement in a book in 1979] there were two concerning religion. I think people need religion, and the notion of the Earth as a living planet is something to which they can obviously relate. At the least, Gaia may turn out to be the first religion to have a testable scientific theory embedded within it."

The "Rights" of Trees

There is something fundamentally wrong with treating
the earth as if it were a business in liquidation.
—Herman Daly

Lynn White argued that belief in the natural world as nothing more than an instrument provided for human use led automatically to its mistreatment and damage. His mistake was in assuming that this link is inevitable. Scientists and nonscientists alike have challenged White about this. Everywhere one finds expressions of awe and respect for the world around us. This may not lead to an understanding of the mechanisms involved in its maintenance, but it is very different from the pervasive arrogance that White saw as the cause of environmental misuse.

A key figure in formalizing a non-anthropocentric attitude to the environment less controversial than that of Naess or Lovelock was John Muir. In the Canadian wilderness north of Lake Huron, he came upon a cluster of orchids so beautiful that he sat down beside them and wept. Reflecting on this experience, Muir realized that his emotion sprang from the fact that the wilderness orchids did not have the slightest relevance to human beings. If he had not seen them, they would have lived and died completely unseen. Nature, he decided, must exist first and foremost for itself and its Creator. Everything had value; the basis of respect for nature was to recognize it as part of a created community to which humans also belonged. Like White, Muir came to see Christianity (and civilization) as obscuring this truth by separating people from nature in a dualistic way. During his life he moved from the Calvinism of his childhood to an attitude close to animism.

Muir exercised enormous influence later in life, largely through essays defending wild places in Alaska and the American West. His emphasis was not of any intrinsic rights for nature, but rather on the beauty and spirituality of wilderness and its contrast to the mean, commercial spirit of the age that would sacrifice such environments to the "Almighty Dollar." He focused his energies on the High Sierra of California; his greatest political triumph was the establishment in 1890 of the Yosemite National Park. Two years later a group of Californians organized themselves around Muir to defend the Park, calling themselves the Sierra Club.

Decades later, the Sierra Club pioneered an attempt to give formal recognition to the value of the natural world. In 1969 the U.S. Forest Service granted permission to Walt Disney Enterprises to build a large ski resort in a high valley in the Sierra Nevada of California. The Sierra Club objected to this proposal. Fearful of the likely result, a Californian lawyer, Christopher Stone, wrote a long essay, "Should Trees Have Standing?" In it, he faced the difficulty that the Sierra Club, in filing the lawsuit, was not recognized as a valid plaintiff. It had no legal right to be heard in court, no "standing" in law.

Stone pointed out that while the club itself would not be harmed by the development, the trees that would have to be removed to make way for the development would certainly be "injured." Who could represent them in the courtroom? Trees and rivers and other natural features obviously cannot institute proceedings on their own behalf. Stone believed that the courts should be sensitive to their need of protection and that society should give "legal rights to forests, oceans, rivers, and other so-called 'natural objects.'" He suggested extending the idea of trusteeship to cover such "natural objects," and so give them "legal standing." Stone contended that damage caused by (for example) polluted air or water could be quantified in the same way as an insurance claim, and that human guardians should be able to collect recompense on behalf of the affected features.

The California Supreme Court rejected the Sierra Club's case by a majority of four to one,[7] but the dissenting judge expressed considerable sympathy for Christopher Stone's argument. He wrote, "The critical question of 'standing' would be simplified and also put neatly in focus if we allowed environmental issues to be litigated . . . in the name of the inanimate object about to be despoiled, defaced or invaded."

The issue of "rights" in nature is contentious and is not relevant in the scientific arena. It remains an area of debate for moralists and lawyers. Its significance in the present context is as an illustration of the shift of attitudes from unquestioned anthropocentrism to at least a modified biocentrism.

The Value of the Natural World

There is sufficiency in the world for man's need
but not for man's greed.
—Mahatma Gandhi

All living things benefit or suffer from other living things. We have already considered some of these interactions—parasitism, predation, competition for food or space or mates. Do we gain any positive advantages from our surroundings?

A moment's reflection shows that we do, in fact, have some very familiar and important benefits, such as food or material for building our homes. Such benefits can be described as "ecosystem services." Giving them a name highlights the positive interactions for us that permeate our involvements with the natural world. They also force us to recognize that natural resources are not infinite, a factor that is becoming increasingly apparent as different benefits come under pressure. We have been slow to learn this.

Ecosystem services can be grouped under four headings:

- *Supporting services*: Primary productivity, nutrient dispersal and cycling, pollination. These underline the other three types of services.
- *Provisioning services*: Food, fuel, energy (fossil fuels, plus hydro and wind power).
- *Regulating services*: Cleansing air currents, purifying water, stabilizing and detoxifying soils, disease control, modifying the climate (not least by removing carbon dioxide from the atmosphere).
- *Cultural services*: Aesthetics, intellectual and spiritual stimulation

The scales on which ecosystem services operate vary from microbes to landscapes, and from milliseconds to millions of years. They include, for example, the detritus on a forest floor, the microorganisms in the soil, and the characteristics of the soil itself, all of which contribute to the abilities of the forest to influence carbon sequestration, water purification, and erosion prevention. Understanding of ecosystem services involves

- Identifying ecosystem service providers (ESPs)—species or populations that provide specific ecosystem services—and characterization of their functional roles and relationships.
- Determining those aspects of community structure that influence how ESPs function in the natural landscape, such as compensatory responses that stabilize function and nonrandom extinction sequences that can erode it.
- Assessment of key environmental (abiotic) factors influencing the provision of services.
- Measuring the scales on which ESPs and their services operate in time and space.

The concept of ecosystem services first surfaced in a 1970 report, *Man's Impact on the Global Environment*, prepared for the first major international conference on the environment in 1970, the United Nations Conference on the Human Environment in Stockholm. It has become increasingly important as the value of such services to humankind has been calculated in monetary terms. A very crude estimate of their magnitude is that the world's ecosystems contribute around $33 million every year to the world's sustenance, a sum twice the annual gross national product of all the nations of the globe. Examples of the connection between ecosystem services and money are:

▶ When the quality of drinking water in New York City fell below the standards required by the U.S. Environmental Protection Agency, the authorities opted to restore the polluted Catskill Watershed that had previously provided the city with the ecosystem service of water purification, rather than constructing a high-tech filtration plant. Once the input of sewage and pesticides to the watershed area was reduced, natural abiotic processes such as soil adsorption and filtration of chemicals, together with biotic recycling via root systems and soil microorganisms, resulted in an improvement in water quality to levels that met government standards. The cost of this investment in natural capital was estimated as a one-off of between $1 billion and $1.5 billion, considerably less then the estimated $6 to 8 billion monetary capital cost of a filtration plant, never mind the $300 million annual operating costs.

▶ Fifteen to 30 percent of U.S. crop production depends on pollination by bees. One study found that in California's agricultural region, wild (i.e., nonhive) bees alone could provide partial or possibly even complete pollination services and would certainly enhance the services provided by honeybees through behavioral interactions. However, intensified agricultural practices can quickly erode pollination services through the loss of species, including bees. This can be compensated for by chaparral and oak-woodland habitat within one to two km (approximately half to one-and-a-half miles) of a farm, thus providing a potential insurance policy for local farmers. Wild bees are likely to become increasingly important given the massive mortality afflicting hive bees.

► Spatial models of water flow through different forest habitats were created in watersheds of the Yangtze River (China) to assess potential contributions for hydroelectric power in the region. By quantifying the relative value of ecological parameters (vegetation-soil-slope complexes), researchers estimated that the annual economic benefit of maintaining forests in the watershed for power services would be 2.2 times greater than if they were harvested once for timber.

► The mineral water company Vittel faced a critical problem in France in the 1980s because of nitrates and pesticides entering the springs that were the source of its water. Intensified agricultural practices and cleared vegetation had allowed impurities to enter the aquifer upon which the company drew. Previously the water had been filtered naturally. This contamination threatened the company's right to label its product "natural mineral water" as defined under French law. In response, Vittel provided subsidies and free technical assistance to farmers in exchange for the latters' agreement to enhance pasture management, reforest catchments, and reduce the use of agrochemicals, thus developing an incentive package for farmers to improve their agricultural practices and consequently reduce pollution.

As our modern world becomes ever more aware of the economic benefits of ecosystem services, the votes and consumer practice of citizens hopefully will support policies that seek a sustainable relationship with nature. Economic incentives are part of this, but they are not enough by themselves. We need to go beyond pure utilitarianism and develop what we can call "environmental literacy."

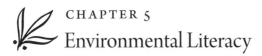

CHAPTER 5

Environmental Literacy

Let it not suffice to be book-learned, to read what others
have written and to take upon trust more falsehood
 than truth, but let us ourselves examine things as we
have opportunity and converse with Nature as well as
with book. . . . I know that a new study at first seems
very vast, intricate and difficult: but after a little resolution
and progress, after a man becomes a little acquainted with it,
his understanding is wonderfully cleared up and enlarged,
the difficulties vanish, and the thing grows easy and familiar.
—John Ray, *The Wisdom of God Manifested*
in the Works of Creation, 1691

HISTORIANS TEND to trace science developing from the observations and theories of astronomers and physicists, from Copernicus to Einstein. But an arguably more important path was one traveled by way of careful observation of the natural world. Such a path can be regarded (and is often denigrated) as nothing more than natural history. Whatever we call it, we should not despise it. The study of nature is much older than formal science. In the fourth century BC, Aristotle was an excellent naturalist, but his observational and experimental approach was sadly neglected for many centuries. Tradition ruled over observation and reality. Then in 1543, the same year of Nicolaus Copernicus's *De Revolutionibus Orbium Coelestium* appeared, a twenty-eight-year-old Belgian, Andreas Vesalius, showed the errors and dangers of relying upon convention in his careful dissections of the human body. In his *De Fabrica Humani Generis* he demonstrated that you must actually dissect the body to see how it works; thereby he laid the grounds for modern scientific rigor.

Accurate observation and recording are an integral part of all science. In the context of our present discussion, they are essential if we are to become environmentally literate. Pioneers such as John Ray and Joseph Banks were excited and stimulated by the natural world; they went out and experienced nature. Ray's first book was on the flora of his local county, Cambridgeshire. He wrote,

> First I was fascinated and then absorbed by the rich spectacle of the meadows in spring-time; then I was filled with wonder and delight by the marvellous shape, colour and structure of individual plants. While my eyes feasted on these sights, my mind too was stimulated. I became inspired with a passion for Botany, and I conceived a burning desire to become proficient in that study, from which I promised myself much innocent pleasure to soothe my solitude.

The same passion fueled Joseph Banks, who went on to introduce the Australian flora to the world and serve as president of the Royal Society of London for thirty years. According to one account, the inspiration began when he was at high school:

> He was walking leisurely along a lane, the sides of which were richly enamelled with flowers; he stopped and looked around, involuntarily exclaiming, How beautiful! After some reflection, he said to himself, it is surely more natural that I should be taught to know all these productions of Nature in preference to Greek and Latin. . . .

Is there any connection between the sort of wonder experienced by Ray and Banks and a desire to protect the world in which we live (the only one we have)? Should not science stand apart from human feelings, protecting its pursuit of objectivity? Can we—or perhaps, even should we—combine scientific analysis with a sensitivity for nature and its future?

To understand how this might be possible, we need to develop and encourage environmental literacy, requiring knowledge as well as sentiments (or experience). Literacy is the ability to read; it necessarily has to incorporate both acquiring information and intellectual discipline. A literate person is one who can understand what is written and place it within a context. Environmental literacy involves head learning from books and libraries but goes further and involves experiencing and interpreting real environments. We can learn from others who have trodden the path in the past, providing us a legacy of ecological research and attitudes.

John Muir was an excellent example of an environmentally literate person. He had little formal learning, but drew inspiration from journeys in the wilderness areas of North America, well expressed in his first book, *The Mountains of California* (1894). Muir's approach to writing about nature was not unlike that of Henry Thoreau, whose account of experiencing life in a wooden cabin appeared in 1854 as *Walden*. Neither Thoreau nor Muir was a scientist. Indeed, Muir was suspicious of the nascent ecology of his time, represented most strongly for him by Gifford Pinchot and the power of the U.S. Forest Service, of which Pinchot was the first chief. Muir and Pinchot fell out over the nature and management of wilderness. Muir valued nature for its spiritual and transcendental characteristics, and any management of it was sacrilege to him. Both Muir and Thoreau were inheritors of a growing tradition of wondering, learning, and increasing understanding of the natural world that had been growing apace since the mid-eighteenth century.

We have already noted some of the drivers in this tradition—

John Ray and the production of useable floras and faunas; the discoveries of deep time and deep space by Hutton, Smith, and Herschel; the natural theology of Paley and the Bridgewater Treatises; the voyaging of Humboldt and the recognition of the need for quantitative studies, followed by the travels of Darwin, Wallace, the entomologist Henry Bates, and Hooker. Can we put the growing knowledge and understanding of natural history represented by these pioneers into a coherent syllabus for environmental literacy?

Identifying all the elements that contributed to the crossover from fear and awe of the natural world in earlier times, to wonder and commitment, and then onto scientific dissection would be difficult. All we can do is point to some of the key players. A convenient starting point is the early sixteenth century, when Copernicus and Vesalius were pioneering in their very different ways. Around this time, five naturalists born within a few years of each other contributed significantly to the revival of biology after the scholastic sterility of the Middle Ages: William Turner (1508–1568), the father of English botany; Pierre Belon (1517–1564) and Guillame Rondel (1507–1566) in France; Ulisse Aldrovandi (1522–1605) in Italy; and most influential of all, Konrad Gesner (1516–1565) in Switzerland, whose four-thousand-page *Historia Animalium* can be taken as the starting point of modern zoology.

Gesner was a true Renaissance man. Alongside his scientific interests, he had strong theological connections. He was born in Zurich. When he was fifteen, his father was killed fighting alongside Zwingli. Gesner knew the Swiss Reformers—Martin Bucer, Heinrich Bullinger, and Conrad Grebel. At one time his life's work seemed to be as a pastor, but he took a medical degree at Basel and settled down to medicine and science. He undertook fieldwork for himself. He wrote,

> I have determined, as long as God gives me life, to ascend
> one or more mountains every year when the plants are

at their best—partly to study them, partly for exercise of body and joy of mind.... I say then that he is no lover of nature who does not esteem high mountains very worthy of profound contemplation. It is no wonder that men have made them the houses of gods, of Pan, and the nymphs....

The century that followed Gesner and his colleagues saw great advances in plant classification, although few in zoology. Progress in zoology was not made until structure replaced function and habitat as the criterion for classification. This step was first made by Francis Willughby (1635–1673) in his posthumously published *Ornithologiae libri tres* (1676), albeit that largely was the work of his friend and editor, John Ray, who we have already met.

With Willughby and Ray, we move from the pedantic scholasticism of medieval Catholic Europe to the increasing intellectual ferment of the Reformation, particularly in Britain. John Ray's *The Wisdom of God Manifested in the Works of the Creation* (1691) knitted together in a seamless way the works of God in creation, redemption, and providence, with God treated as a benevolent dynamo rather than a distant mechanic or a Cartesian *deus ex machina*. One hundred and fifty years later Paley plagiarized Ray without acknowledgment, and wrote of a creation running under God's governance but largely independent of him. The Reformation detached God from the clutches of the ecclesiastical hierarchy, but the Enlightenment that followed sent the deity back to the beginnings of time, relegating him to little more than a first cause.

As noted earlier, Paley massively influenced attitudes in the nineteenth century, but a growing corpus of knowledge was undermining his version of natural theology. The eighteenth century saw Buffon and Linnaeus laying the framework of biogeography as more of the world became known. James Cook's circumglobal voyages (particularly the first two, in 1768–1771 and 1772–1775) can be regarded as the first major expeditions with a large scientific component.[1]

While Cook and Banks were sailing in the southern ocean, a somewhat self-important Welsh landowner, Thomas Pennant (1726–1798), took part in a couple of cruises to the west coast of Scotland. He published the account of his journeyings as *A Tour of Scotland 1769* with a further volume a few years later. These proved very popular and made Pennant's reputation. However, he was essentially an intellectual entrepreneur and popularizer who drew more on secondhand information than firsthand description. His method of obtaining information was chiefly through answers he collected from an army of correspondents. Nonetheless, he had a good eye for scenery, and his travel books were best-sellers. Pennant can, perhaps, be regarded as a proto-ecotourist. The relevance here is that one of his correspondents was Gilbert White, an Anglican clergyman in central southern England.

White wrote a series of letters to Pennant detailing his observations in and around his parish, but became increasingly disenchanted with Pennant's demands on him. His response was to gather together his correspondence with Pennant with some other documents and publish them himself in 1789 as *The Natural History of Selborne*. White's *Selborne* has remained in print ever since and influenced generations to follow in his footsteps in observing and recording the natural world. To take but one example, Charles Darwin wrote about his time at school, "From reading White's *Selborne* I took much pleasure in watching the habits of birds, and even made notes on the subject. In my simplicity I remember wondering why every gentleman did not become an ornithologist."

White was a talented and assiduous natural historian, as well as unquestioning in his acceptance of the Christian God as a good and omnipotent Creator. But he lived at a time of turmoil. The year that saw the appearance of *The Natural History of Selborne* was also the year of the storming of the Bastille in Paris and the start of revolutionary disruption throughout Europe. It was also the height of the so-called Enlightenment, which developed more or less simultaneously in many countries and which placed reason as the primary source and legitimacy for authority. The Enlightenment was a challenge to received ideas of power. But in addition it was a coming of age of the stumblings toward modern science begun by Nicolaus Copernicus, Andeas Vesalius, and Galileo Galilei. It was a time characterized by critical questioning of institutions, customs, and morals.

A key Enlightenment concept was "Nature." For early Enlightenment thinkers, Nature linked the divine (eternal and transcendental) with the human; it pointed to the purification and perfectability of humankind, and effectively deified the natural world. Such thinking greatly influenced the Industrial Revolution on the one hand and Romanticism on the other. In Britain, the eighteenth century flowered as the first great age of landscape painting and aesthetic writing.

The complexities and tensions of the movement from Enlightenment to Romanticism are well illustrated by Samuel Taylor Coleridge (1772–1834), best known as the friend and fellow Lake District poet of William Wordsworth. But Coleridge was much more than a Romantic poet. He was a supporter and correspondent of Sir Humphry Davy, chemist and president of the Royal Society of London from 1820 to 1827. Coleridge attended and was lionized at a historic meeting of the British Association for the Advancement of Science in 1833 where the word "scientist" first replaced the term "natural philosopher." At the meeting, Coleridge denounced the French rationalist philosopher René Descartes and came close to acclaiming Charles Lyell's geological uniformitarianism. He wrote elsewhere on the nature of consciousness and seemed near to accepting that humans had evolved. Coleridge well illustrates the contradictions and closeness between science and aesthetic involvement in the early nineteenth century.

The years at the end of the eighteenth century saw social and technological upheaval as well as political and intellectual revolution.

They were also marked by a group of pioneer industrialists and local leaders who met informally in the English Midlands for almost fifty years, from 1765. Their meetings were held at the time of the full moon, so that they had light to help them on the way home after their meeting; for this reason, they called themselves the "Lunar Society." The group included the steam and railway engineer James Watt, the potter Josiah Wedgwood, the clockmaker Jonathan Whitehurst, the chemist and Unitarian preacher Joseph Priestly, Charles Darwin's grandfather Erasmus, the manufacturer Matthew Boulton, and William Small, former professor of natural philosophy at the College of William and Mary in Williamsburg, Virginia. They were occasionally joined by Benjamin Franklin, when he was in England doing business for the American colonies. The discussions of the Lunar Society mirrored the events of the Industrial Revolution in Britain, with its increasing urbanization and consequent changes to the countryside.

In 1771 Joseph Banks returned to England from Cook's first voyage and introduced the plants of the Southern Hemisphere to Europe. Then in 1788 James Hutton's "Theory of the Earth" was published in the *Transactions of the Royal Society of Edinburgh*, spelling out the beginning of modern geology and the beginning of the end of Neptunism—the belief that all rocks had been precipitated out of a global flood. From 1799 to 1804 Alexander von Humboldt was traveling in Latin America, making observations that formed the basis of quantitative ecology, physical geography, and meteorology, and inspiring a generation of naturalists—including Darwin, Wallace, and Bates. On his way back to Europe, Humboldt stayed with Thomas Jefferson in Washington at almost the same time that the president sent out Meriwether Lewis and William Clark from Missouri to explore the land the French government had sold to the United States. They were charged with recording the geology and botany of the country through which they passed, as well as the human challenges of the West.

While Lewis and Clark were discovering wilderness in America and only five years after *The Natural History of Selborne* appeared, William Paley published his *View of the Evidences of Christianity*, a work that influenced attitudes about the natural world for more than half a century. All this turmoil catalyzed a general—almost a passionate—interest in collecting and exhibiting specimens of the natural world, particularly in Britain, an interest fostered and enabled by the increasing freedom of travel as railways spread and better roads made travel easier. From this fascination of men and women with the natural world, the science of ecology sprang.

We have already seen that the first ecological society in the world came from capitalizing on the passion for collecting wildflowers (p. 7). By 1900 there were around one hundred thousand members of local societies in Britain, many of which had "botanizing" as their chief activity. As time went on, they became increasingly involved in mapping local plant communities, and it was this that led to the formation of the Committee for the Survey and Study of the British Vegetation, which within ten years became the founding core of the British Ecological Society.

Parallel developments took place in the United States, most famously a botanical survey of Nebraska by Roscoe Pound, who went on to be dean of the Harvard Law School, and Frederic Clements, who became one of the most influential botanists in America. Zoologists began to undertake similar surveys. Charles Davenport (1866–1944), later notorious for his promotion of eugenics, conducted a study of the animals on the Cold Spring Sand Spit on Long Island, New York, just down the road from the laboratory established by the Carnegie Institute. He was followed by Victor Shelford in Illinois, who famously found the task of organizing ecological data to "seem hopeless."

Shelford summarized much of his early work in *Animal Communities in Temperate America as Illustrated in the Chicago Region*, published in 1913. Julian Huxley (grandson of Thomas Henry Huxley and younger brother of Aldous Huxley), teaching in the Zoology Department at Oxford University, gave a copy to a first-year student Charles Elton, who took it with him when he accompanied Huxley as a field assistant on a biological expedition to the Arctic island of Spitzbergen in 1921.

Shelford had concluded that it was possible to define the animal communities of different habitats by the physiological reactions of "indicator species" confined to a narrow range of habitats. On Spitzbergen, Elton found Shelford's method of distinguishing different animal communities to be inadequate. There were very few species in the Spitzbergen communities, and even fewer confined to a single habitat or plant association. He found it much easier to identify animal communities in terms of their food relationships. Elton wrote,

> Food is extremely scarce in the Arctic, both on land and in fresh water, though it is plentiful in the sea. Most of the scavenging animals live on decaying animals and are therefore practically equivalent to herbivores. There are no elaborate "chains" of species which depend on animals which eat the dung or decaying bodies of other species. Such a "short-circuiting" of the nitrogen cycle (which exists in other countries, *e.g.* badgers eating beetles which prey on dung beetles in England) appears to be unimportant in Spitzbergen. Dead animals are very rarely found, and when they do occur are devoured by vertebrates (*e.g.* reindeer by Glaucous Gulls and dead whales by bears). Where animals like bears have the chance of decaying, they do so very slowly.

Elton acted as chief scientist on further expeditions to Spitzbergen in 1923 and 1924, which gave him the opportunity to look closely at numerous habitats (see Figure 5.1). He found only nine species of animals among a few species of lichens and mosses on the eastern side of Spitzbergen, whereas he recorded over fifty species in the heaths of the western side. The complexity of the animal

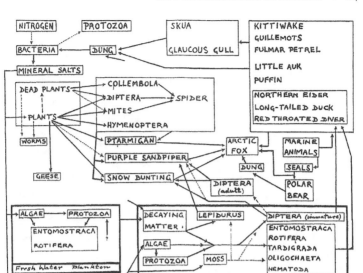

Figure 5.1. Elton's diagram of the nitrogen cycle on Bear Island (from Victor S. Summerhayes and C. S. Elton, "Contributions to the Ecology of Spitsbergen and Bear Island," *Journal of Ecology* 9 [1923]: 214–86, by permission of Wiley-Blackwell).

and plant communities varied according to the severity of the climate. He declared that the chief scientific goal of pure ecology was the way in which population numbers were regulated.

On his way back from Spitzbergen in 1923 Elton bought a copy of Robert Collett's book on Norwegian mammals (*Norges Pattedyr*). It introduced him to the four-year cycle of lemming abundance. Back in Oxford, Huxley drew Elton's attention to another book (*The Conservation of the Wildlife of Canada*, by Gordon Hewitt); Elton was intrigued by a chapter on the periodic fluctuations in the populations of fur-bearing animals, based on the returns of fur trappers from the Hudson's Bay Company. At the time, a belief in some sort of harmony—a balance of nature—in natural communities prevailed. Such harmony implied that any tendency of a species to increase in numbers would be cancelled out by some controlling factor. Elton argued that no evidence of any such balance existed.

Elton distilled his conclusions in a remarkable two-hundred-page book, *Animal Ecology*, published in 1927. Commissioned by Julian Huxley, Elton wrote the book in eighty-five days. In it he set out the key features of animal communities, including their characteristic species, zonation, and species richness. He described his approach as dealing with the "sociology and economics of animals" rather than their structural and other adaptations. He pointed out that by studying the differences between habitats, we can often identify the fundamental resemblances between them. Elton elaborated the concept of the niche to describe an animal's place in the community—"what it is doing, most specifically its relations to food and enemies" (p. 44)—and invoked the roles of time and weather as affecting the segregation of niches. He coined the term "pyramid of numbers" to describe the relative decrease in numbers at each stage in the food chain. He stressed the role of enemies in the regulation of numbers. And all this from a quiet and retiring man a few years after his undergraduate degree.

For Elton, ecology meant scientific natural history. His experience on Spitzbergen with its few species and extreme environments enabled him to recognize some of the essential processes affecting and controlling natural populations which are overlain and therefore obscured by the larger number of species and greater complexity of habitats in more temperate areas . For him, ecology had become scattered, like "an active worm that has been chopped into little bits, each admirably brisk, but leading a rather exclusive and lonely existence and not combining to get anywhere in particular." Spitzbergen changed that. Elton's niche concept was useful because it was so simple. The Berkeley zoologist Joseph Grinnell (1877–1939) had earlier defined niche in terms of competition, but Elton regarded competition as not particularly important, although he later came to accept that it could provide some clues to community organization.

Modern science demands quantitative rigor. Following Grinnell and Elton, a group of more theoretically inclined scholars refined and developed the concept of the niche by turning it into a multidimensional idea, although in doing so they often obscure its clarity. Elton was certainly environmentally literate, but he was not a mathematician. Peter Crowcroft, one of Elton's students, commented, "The initiative of ecologists skilled in algebra has provided material for many theses and publications, but has advanced knowledge of the natural world about as much as theological databases about the size and morphology of angels."

As a consequence of reading Hewitt's *Wildlife of Canada*, Elton became fascinated by the fluctuations in numbers of fur-bearing animals and this led to his appointment as biological consultant to the Hudson's Bay Company. In this capacity, he acted as secretary of a conference on biological cycles, held at Metamek, a remote village on the Gulf of St. Lawrence in 1931.

University of Wisconsin ecologist Aldo Leopold was also at Metamek. At the time Leopold was employed by an industry consortium, the Sporting, Arms, and Ammunition Manufacturers' Institute, to carry out a series of game surveys. Leopold had trained as a forester and knew just enough theory to give his observations credence. He was a manager, imbued with ecological ideas. He saw conservation as "a state of harmony between man and the land." In 1931, Elton was laying down the foundations of ecology, while Leopold was attempting to apply the science before its foundations were properly set. They became friends and mutually influenced each other. Elton was persuaded of the effect of humans on animal population numbers while Leopold realized the need for good ecological data to understand how numbers varied. Elton's writing increasingly bore the mark of Leopold's conservation thinking.

An older contemporary of Elton and for many years a colleague at Oxford University was Arthur Tansley, whom we have already met in his criticizing of Clements and attempt to demystify illusions about the natural world. Tansley's ecosystem concept was enthusiastically embraced by two brothers, Eugene and Howard Odum, who through Eugene's textbook (*Fundamentals of Ecology,*

1953) influenced generations of students, particularly in North America. Tansley dominated botany for almost the whole of the twentieth century, especially in Britain, through his editorship of influential journals and his increasing involvement in conservation issues as he saw the damage being done to the places he studied and loved. Tansley's last publication was a pamphlet, *What Is Ecology?* extolling the way in which the science of ecology brought together the understanding of investigable mechanisms with the marvelous diversity of the natural world.[2]

All this brings us back to environmental literacy. Jean-Jacques Rousseau, one of the intellectual inspirers of the French and American revolutions, used to walk a mile from his home solely for the pleasure of hearing a nightingale. Shakespeare's Falstaff died "babbling of green fields." Thoreau tells us "there are moments when all anxiety and stated toil are becalmed in the infinite leisure and repose of nature." Julian Huxley, archetypal agnostic humanist, wrote of gazing at the stars and feeling he could possess their immensity: "The joy of it filled my heart like a revelation, a reassurance that the world of natural beauty meant something to me, and to the world."

Is it possible to analyze this passion for the natural in any rational way? We can profess a distaste for the "unnatural" like nuclear power or ribbon developments along highways, but behave as if washing machines, e-mails, and cell phones are essentials. Although we may truly envy people who live a simple life and profess a willingness to do without television and flush toilets, we still want modern medicines and access to rescue services when we are lost. Thor Heyerdahl of *Kon-Tiki* and *Ra* fame spent a miserable year with his newly married wife seeking Utopia on a remote island, one of the Marquesas group in the South Pacific. On his return to Europe he wrote, "There is no Paradise to be found on Earth today. There are people living in great cities who are far happier than the majority of those living in the South Seas. Happiness comes from within, we realise that now. . . . It is in his mind and way of life that man may find his Paradise—the ability to perceive the true values of life, which are far removed from property and riches, or from power and renown."

The progression from Elton—a naturalist par excellence—to Heyerdahl, a rationalist philosopher, via Leopold the pragmatic molder of the natural world, seems to take us far from scientific analyses and ecology proper, yet there is an important connection. Leopold linked the ecological science of Elton with problems of ethics, morality, and aesthetics, without downgrading the need for data on science, resource management, and public policy. Does the study of ecology necessarily lead to introspection and ethics? There are purists who would regard such a progression as wholly illegitimate, but this is only possible if we take a very narrow view. The reality is that humans exercise such a massive influence on the natural world that it is eccentric to ignore it. The breadth of this influence is highlighted by the different understandings of Elton and Leopold. Elton did not believe that there was such a thing as a "balance of nature." In contrast, Leopold was a firm believer in it. He viewed animal populations as being in a state of dynamic equilibrium maintained by environmental resistance inhibiting biotic potential; he treated land as resembling an organism sometimes disturbed from a state of "aboriginal health." In trying to reconcile such tensions and determine if there are proper limits to the study of ecology, it is proper to probe the ecology of humans. That requires enquiring into the nature of the human animal.

CHAPTER 6
The Proper Study of Mankind

WHERE DO HUMANS fit into the natural world? Are we apes on the way up? Angels on the way down? Pilgrims on a journey or flotsam in a flood? Are we embodied souls or mere DNA reproducing machines driven by deterministic physico-chemical reactions? The nature of humanness is the meeting point for anthropology, psychology, and sociology—sciences often in tension with how human nature is defined by theology. Is this important—or relevant—for ecology? The English poet Alexander Pope pointed to the tension between science and theology when he enquired where humans stood in the wider world—are we somewhere between God and beast, he asked. He answered his own question by urging that "the proper study of Mankind is Man."

Our conclusions about humanness must be tentative (and we shall wait until the final chapter to probe the science-or-theology question). But any definition of being "human" must be grounded in the human body itself—a mass of flesh begun by the fusion of two genome-carrying gametes, nurtured in a womb for nine months or so, and modified by a gamut of influences—a biological entity that is an "it" as well as an "I." The study of the human organism and its reactions in a range of environments is not so different from the study of ecology. Not a few modern thinkers have drawn parallels between sociology and ecology and, as we shall see, the question of human ethics and altruism has produced a theory called sociobiology to offer some tentative explanations. The difficult questions arise when we try to define in what ways we are completely different from other animals.

Scholars through the ages have wrestled to define the character-istics of humanness. The assumption used to be that humans alone in the animal kingdom can reason and use tools. The trouble is that animals, too, can learn and may be able to solve problems that require a degree of intelligence. Moreover, examples of tool-using animals are increasingly being found—such as birds using twigs to tease out insects from cracks. Our old definitions of humanness are obviously inadequate; they had to be completely abandoned when chimpanzees were seen selecting and then modifying their tools. They were toolmakers as well as tool-users.

It is easy to get diverted in any search for unique human attri-butes. Most debates about humanness—certainly the most heated ones—have tended to be deflected into controversies about our origins and antiquity. There can be no disagreement that we are animals, members of a single species, *Homo sapiens*. What is our relationship to other animals? Did we we somehow originate by something unique and distinct from nature, perhaps a "soul"? The problem here is that the understanding of the soul is becoming as awkward as the old definition of humans as the only creatures able to reason. For almost two millennia most people assumed that we are made up of body and soul, or perhaps body, soul, and spirit; theological libraries groan under competing biblical justifi-cations for a bipartite versus a tripartite nature. But these discus-sions are proving irrelevant; the idea of a transcendent soul can no longer be sustained. Neuropsychological research emphasizes how human beings are a psychophysical unity.[1] This is not to deny that some philosophers still claim to be able to combine the idea of psy-chophysical unity with a sophisticated version of dualism, but their arguments are less and less convincing.

On the question of our antiquity, it was clear by the end of the eighteenth century that life had been on Earth for a very long time, though human origins were still assumed to have arrived around six thousand years ago, since this was when the biblical genealogies that go back to Adam begin. However, even this was not as straight-forward as it seemed. We have already met the French Jew, Isaac de

La Peyrère who argued that the biblical Adam and Eve were not the original human beings, but merely the first parents of the Jewish people. For him, there were "pre-Adamites," which helped explain how Cain found a wife and who had peopled the land of Nod. La Peyrère was heavily criticized in his time, but the notion of pre-Adamites persisted. It made sense to a lot of people puzzling over our antiquity. A complication was that it implied "polygenism," the possibility that there were several human origins, God creating separate races in different lands.

It is said that one of Darwin's motives in developing his ideas of "descent with modification" was to establish as firmly as possible that all human beings are part of the same stock, that is, that we are not divided into different species with different statuses, as some of his contemporaries claimed, most notoriously the Harvard zoologist Louis Agassiz (1807–1873). Nevertheless, the debate about species or races rumbled on for more than a century. For Christians it was complicated by the publication (and general acceptance) of Darwin's ideas about our animal origins plus evidence appearing around the same time from the excavation of Kents Cavern (near Torquay in western England), that humans once lived among now-extinct creatures, and the discovery in 1856 of fossils in the Neander Valley near Düsseldorf of an apparently primitive form of human, soon named Neanderthals.[2]

To track human origins, we need to concentrate on the *hominid* family, those mammals who move on two legs, and then within the family, the genus called *Homo*, of which modern humans are a part. Since the first Neanderthals were found, many hominid fossils have been discovered, to the extent that it is fair to claim that *Homo* has a better fossil record than almost any other genus of organisms on Earth. Unfortunately the credibility of these hominid discoveries has often been marred in popular misunderstanding by overimaginative reconstructions. There have been repeated fanciful attempts to portray human ancestors as either hulking brutes or mere variants of modern individuals. Even today, the image of human fossil history for many is probably the much-reproduced frontispiece of T. H. Huxley's *Man's Place in Nature*, published in 1863. It shows a parade of modern skeletons from a gibbon, through a series of stooping apes, to an upright man ("A grim and grotesque procession," as Darwin's critic, the Eighth Duke of Argyll, called it), implying a direct link between living primates and modern humans. A more accurate picture would be a bush, with many branches and twigs that seem to lead nowhere, and modern apes and humans sharing an ancestor near the trunk. Only one of these branches gave rise to the various species within the *Homo* genus and only one from there on led to modern humans.

HUMAN ORIGINS

The oldest known fossil hominids are 6 to 7 million years old and come from Africa. They are known as *Sahelanthropus* and *Orrorin* (or *Praeanthropus*). These putative ancestors tended to walk on two feet when on the ground and had very small brains. A more recent fossil is *Ardipithecus*, who lived about 4.4 million years ago in Africa. Following on, there are numerous fossil remains from many African sites of *Australopithecus*, a hominid that appeared between 3 million and 4 million years ago. *Australopithecus* had an upright stance like modern humans but a brain size (more accurately, a cranial capacity) of less than 500 cubic centimeters, about the same as a gorilla or chimpanzee, but one-third that of our brain. The skull of *Australopithecus* had a mixture of ape and human characteristics— low forehead and long, apelike face, but with teeth proportioned like those of humans. Other early hominids partly contemporaneous with *Australopithecus* include *Kenyanthropus* and *Paranthropus*; both had smaller brains, although some *Paranthropus* had larger bodies. *Paranthropus* is an example of a side branch of the hominid lineage that became extinct.

More human characteristics appeared in *Homo habilis*, which lived between about 2 and 1.5 million years ago in Africa. It had a cranial capacity of six hundred to seven hundred cubic centimeters, only slightly larger than that of *Australopithecus*. *Homo erectus*, which appeared in Africa somewhat before 1.8 million years ago, had a cranial capacity of eight hundred to eleven hundred cubic centimeters.

Homo erectus was the first intercontinental wanderer of our hominid ancestors. Fairly soon after its emergence in Africa, *H. erectus* spread to Europe and Asia. Fossil remains have been found in Africa, Indonesia (Java), China, the Middle East, and Europe. *Homo erectus* fossils from Java have been dated at 1.81 and 1.66 million years ago, and from Georgia (in Europe near the Asian border) between 1.8 and 1.6 million years ago.

Several species of the genus *Homo* lived in Africa, Europe, and Asia between 1.8 million and 500,000 years ago. They are known as *Homo ergaster*, *Homo antecessor*, and *Homo heidelbergensis*, with brain sizes similar to that of the brain of *Homo erectus*. Some of these species overlapped in time, though they lived in different regions of the Old World. They are sometimes lumped together under the name *Homo erectus*. The transition from *Homo erectus* to *Homo sapiens* may have started around 400,000 years ago. Some fossils of that time appear to be "archaic" forms of *H. sapiens*. Despite its eventual demise, *H. erectus* persisted until 250,000 years ago in China and perhaps until 100,000 years ago in Java. (The fossil remains of *Homo floresiensis*, discovered in 2004 on the Indonesian island of Flores, seem related to *H. erectus*, although *H. floresiensis* was much smaller and lived around 12,000 to 18,000 years ago. These fossil remains are being actively investigated; their precise identification remains controversial.)

We then come to the Neanderthals, the first great fossil finds to excite the public imagination. The hominid branch of *Homo neanderthalensis* appeared in Europe more than 200,000 years ago and persisted until 30,000 years ago. It was once thought the Neanderthals were our direct ancestors, but we know now that modern humans appeared more than 100,000 years ago, long before the disappearance of Neanderthals. The relationship between Neanderthals and modern humans is certainly complicated. In some Middle East caves, modern humans precede and also follow Neanderthals. Modern humans found in these caves have been dated at 120,000 to 100,000 years ago, Neanderthals at 60,000 and 70,000 years, and then more modern humans dated at 40,000 years. It is unclear whether Neanderthals and modern humans replaced one another by migration from other regions, or whether they coexisted. Comparisons of DNA from Neanderthal fossils with living humans seem to show that some interbreeding occurred between Neanderthals and their contemporary, anatomically modern humans.

Fossils of the genus *Homo* show fairly consistent trends in increasing brain and decreasing tooth size. Often the changes were rapid. Between early *Australopithecus* and early *H. erectus*, brain size increased by about 50 percent and then by another 50 percent again between *H. erectus* and the present. Assuming that *Homo* and *Australopithecus* shared a common ancestor, this growth must be a genetic difference (not just an environmental effect). The second period was only half as long as the first. It has been calculated that the second phase of increase, though dramatic, could be achieved by natural selection (a *selection differential*) of only 0.00004 of each generation. This is very small compared to other selection pressures operating in the wild (10,000 times less than that observed in the Galapagos Islands finches, for example). Notwithstanding, the rate of increase of human brain size is among the most rapid of known evolutionary processes.

Knowledge of the universality of the genetic code and the extraordinary (and unexpected) amount of continuity in coding sequences among very different organisms make it virtually certain—effectively beyond reasonable doubt—that all primates, mammals, and vertebrates have a common ancestry. With the advent of rapid DNA sequencing methodologies, molecular phylogenetics has advanced from the crude comparison of a few large molecules in different organisms to a science capable of identifying quantitative differences between primary gene products and the timing of these in relation to assumed branching events.

For example, many of the genes concerned with smell are inactive in humans but in none of the other primates; the same molecular change in humans, chimps, and gorillas means they all lack a functioning α-fucosyltransferase enzyme, whereas orangutans and monkeys have a functional version of the enzyme (a protein that catalyzes chemical reactions); humans along with the apes and Old World monkeys all have a cytochrome b gene, which is not possessed by the New World. Many more such examples are given in the standard textbooks of biochemical genetics. They can be used to construct robust pedigrees, much more certain than those based on morphological features, not least because anatomical traits can have different genetic bases, and therefore their use in pedigrees is inevitably equivocal.

The findings of genetics and molecular biology are linking us ever closer to other primates. An earlier calculation that humans and chimps share 98.4 percent of their genes has been refined, now that both genomes have been sequenced; nevertheless there is a clear match between more than 95 percent of the genes in the two species. This 5 percent difference amounts to around ten thousand nucleotide changes (individual chemicals that make up the gene), most of them in so-called junk DNA (misnamed because we now know that it has an important function in regulating the activity of protein coding sequences). The high proportion of genes we share with chimpanzees has led to the suggestion that we would be classified as a third chimpanzee species if normal taxonomic criteria were applied.

Our genes are carried on two sets of thread-like chromosomes (one set inherited from each parent). We have one less pair of chromosomes than all other apes (twenty-three pairs instead of twenty-four), but the difference is the simple result of an end-to-end fusion between two separate elements of the ape chromosome set. Most of the chromosomal differences between humans and the chimpanzee are what geneticists call small random duplications; other differences are small insertions or deletions.

There do not seem to be many genes in the human genome that are not found in other apes. Many of our genes are inserted viruses (so-called retroviral inserts), which influence gene regulation in significant ways. There does not seem to be any evidence of positive selection for genes concerned with brain development or function as a whole between chimps and humans. However, one of the most intriguing results from comparative molecular genetics is that since the human and chimpanzee lines separated, one-third more chimpanzee genes show signs of selection than do human ones. Put another way, the implication is that chimpanzees are more specialized than humans. We are generalists, while they are adapted to a particular niche. There is no scientific support for the notion that we have been propelled toward a predetermined end by a Divine Watchmaker or even a Blind one.

BIOLOGY AND BRAINS

For all the genetic similarities, there are obviously enormous differences between us and the apes. We can learn a lot from them. An obvious and important difference is in development rates. Although chimpanzees and humans are in the womb for similar periods, humans take twice as long to mature after birth. For example, our first teeth appear around six months after birth, and our permanent teeth at about six years as compared with three months and three years in chimpanzees. Humans grow for twenty years and live for seventy years, compared to ten and thirty-five years in the apes. Sexual maturity in human females takes place at twelve years of age, but at six to eight years in the apes. These differences must be inherited; they are due to changes in genes, or mutations. Mutations producing similar "neotenous" changes—changes in growth or maturation rates—are known in many organisms (see Figure 6.1).

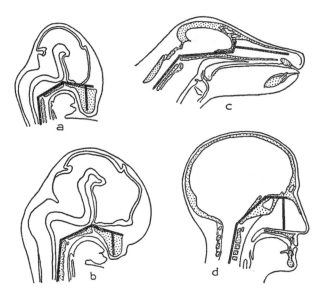

FIGURE 6.1. Neoteny: The adult human retains some of the characteristics of an embryo. The heavy line shows the angle that the head makes with the trunk. In the adult dog (c), the angles are very different from those in the embryonic dog (a), but the angles remain unchanged in the adult human (d) from the embryonic situation (b) (from Gavin R. De Beer, *Embryos and Ancestors* [Oxford: Oxford University Press, 1958], after L. Bolk, *Das Problem der Menschwerdung* [Jena, 1926], reproduced by permission of Oxford University Press).

The large brain of humans is only possible because it can continue to grow after the constraint of having to pass through the mother's pelvis during birth. The human skull is not ossified (i.e., hardened to bone) at this stage. This allows the human brain to continue to increase in size, which it does for a year or so compared to only a month or two in chimpanzees. However, this brings a new set of problems. Large brains require large amounts of energy. About 20 percent of our energy is needed to support brain metabolism. This compares with 9 percent in chimpanzees and 2 percent in marsupials. Furthermore, large brains have a low tolerance for variations in changes in temperature, blood pressure, and oxygen supply, which means there must have been an acute pressure (and therefore strong selection) for our ancestors to find ways of increasing their energy intake. The likelihood is that a shift from vegetarianism to eating meat became important for the survival of a large-brained line. By eating animals (and not just fruits and grain), hominids could ingest more calories. Archaeology bears this out.

Sites where early *Homo* fossils are found often also have animal bones and stone tools. This is probably significant, as is the reduction at the same time of both the protruding (*prognathous*) lower jaw and size of the molar teeth. These changes are consistent with less need to eat tough and fibrous foods. In turn, hunting probably encouraged closer social grouping for defense, food storage, and male-female bonding for increased stability. Studies of fossil skull size and tooth eruption order indicate that the change in growth pattern took place in *Homo erectus* about a million years ago.

Language must have become increasingly important in the human line. The complexity of language is a unique human characteristic. We can assert this without denying the sophistication of communication in many nonhuman groups. Indeed, it can be argued that *the* key difference between chimpanzees and us is our ability to communicate in a spoken language with a large vocabulary, while the apes are apparently incapable of uttering vowel sounds. Every creature can make sounds, but only the human can control these precisely. The essential stage in human differentiation seems to have involved the structure of the larynx, tongue, and associated structures. The supra-laryngeal pathway, the route by which air passes from the nose or mouth to the lungs, is entirely different in humans from all other mammals. In humans, the pathway acts as an acoustic filter. It cannot perform this function in apes and monkeys because their tongues are contained entirely in their mouths. Newborn humans are like this until about three months of age, when over a period of about nine months the mouth migrates backward relative to the base of the skull. The missing ingredient that prevented the chimps developing more complex speech may have been an apparently minor modification of the proto-human vocal tract, which allowed us finer control and permitted the production of a much greater variety of sounds. In other words, only a tiny change in anatomy long ago might be what prevented chimps from "talking."

We do not know for certain when humans diverged from the apes in this respect; it was probably in the early history of the human lineage. However, between 10,000 and 100,000 years ago, two nucleotide differences appeared in the human line in a gene called *FOXP2*, which affects grammar, speech production, nonverbal intelligence, and nonspeech-related movement of the mouth and face, plus cerebellar development. This gene is highly conserved across history: mice and primates differ in only one out of its 715 amino acids (which are produced by the DNA code and which, in turn, make up proteins, the building blocks and catalysts of biological life).

Support for the importance of the *FOXP2* gene in language comes from a family in which fifteen out of thirty-one individuals in three generations of one family have a mutation on the long arm of chromosome 7, which produces a complex of symptoms, including being unable to speak intelligibly. As it turns out, the mutation on the same gene in mice disrupts their ability to produce ultrasonic sounds as well as producing defects in their immature brain.

It is uncertain when humanness (however defined) first arose. Claims have been made that it is shown by formalized burials (implying a belief in life after death), beads and body ornaments, cave paintings, or the origins of agriculture and animal domestication. Many of these components of "humanness" can be recognized in the African archaeological record forty thousand to fifty thousand years ago (stone tools, specialized hunting, use of aquatic resources, long-distance trade, art and decoration). These factors could suggest that humanness developed over a long period.

At the same time, none of these discoveries undermines the crucial role that the development of language must have played in humanization. Language facilitated social intelligence, which in turn involves at least six different faculties: abstract thought; the ability to cooperate in forward planning; problem solving through behavioral, economic, and technological innovation; "imagined communities"; symbolic thinking; and a "theory of mind." A key feature of such a theory of mind is the ability to understand that other individuals may have ideas and desires different from one's own. Partial answers to some of these uncertainties blur the behavioral distinctiveness of humans while recognizing that the differences between the apes and us are so great that they may be seen as qualitative. It could be that the appearance of humanness has been a gradual process over a long period, with one of the keys—the appearance of language—precipitating it only relatively recently, perhaps in the past forty thousand years or even later.

Some anthropologists have argued that the transition from *Homo erectus* to archaic *H. sapiens*, and later to anatomically modern humans, occurred concurrently in various parts of the Old World (Africa, Asia, and perhaps Europe). However, analyses of DNA from living humans has confirmed the African origin of modern *H. sapiens*, and most scientists believe that modern humans first arose in East Africa somewhat earlier than one hundred thousand years ago and spread from there throughout the world, replacing the existing populations of *H. erectus* and related hominid species, including *H. neanderthalensis*. Southeast Asia and the region that is now China were colonized by sixty thousand years ago and Australasia at about the same time. Europe was occupied later, only about thirty-five thousand years ago, and America even more recently, perhaps only fifteen thousand years ago.

Why modern humans reached Europe later than they reached China and Australasia, which are geographically much more distant, is not known. Possibly the Neanderthals, who occupied much of Europe between two hundred thousand and thirty-five thousand years ago, impeded their colonization. Some proponents of the African replacement model suggest that there may have been only a very small band of individuals who were the parents of this global diaspora, narrowing down the direct ancestors of modern humans to a very few individuals. However, this bottleneck in numbers was fairly wide: molecular studies indicate that it involved not less than about ten thousand people and probably nearer a hundred thousand.

Ethnic differentiation between modern human populations is recent in evolutionary history. It seems to be the result of divergent evolution between geographically separated populations during the last fifty to one hundred thousand years.

HUMAN NATURE

The codiscoverer of natural selection with Charles Darwin, Alfred Russel Wallace, argued that natural selection had acted in the earlier stages of human differentiation from the apes, but as our intellectual and moral faculties became "fairly developed," the body ceased to be subject to selection, and subsequent adaptation was solely "through the action of the mind." Wallace's starting point was a belief that brain size was a reliable indicator of mental ability. The difficulty as he saw it was that both fossil humans and "savages" (his word) had skulls (and therefore brains) of similar size to those of civilized people, and consequently all must be presumed to have the same mental capacities. However, it seemed to Wallace that such traits as mathematical ability and the ability to carry out complex trains of abstract reasoning would be useless (if not harmful) in the struggle for existence in primitive cultures. As it was both unneeded and unused, these properties of the human brain could not have evolved by natural selection alone. Consequently, and certainly influenced by his belief in spiritualism, Wallace proposed that a "Higher Intelligence" had guided human evolution in "definite directions and for special ends."

Darwin disagreed with Wallace's conclusion that "natural selection could not have done it all," but he was himself unsure how selection might have acted to produce morality. He wrote in *The*

Descent of Man, "He who was ready to sacrifice his life, as many a savage has been, rather than betray his colleagues, would often leave no offspring to inherit his noble nature." A possible answer to this problem would be if a group rather than an individual is the target of selection. This is one reason that proposals for such "group selection" are frequently advanced. In effect, group selection is very similar from the assumption of superorganisms made in the early days of ecology by Frederic Clements (see p. 28). But quite apart from the original criticisms of Clementsian ideas, group selection has remained highly controversial because of all we know about biological fitness and gene transmission.

In debates on this topic, much effort has been devoted to exploring if there was any possibility it could be in an individual's genetic self-interest to protect its group, even to die for it. As we have seen (p. 29), it was J. B. S. Haldane who first showed a possible way forward. He proposed that individuals could, in effect, inherit genes that made them unselfish, if their effect was to enable them to help the successful breeding of near relatives. Such genes would reduce their possessor's chance of reproduction, so they would only persist in a population if they helped their possessor to support others with the same genes. We might call these "altruistic genes." Haldane calculated that such genes need not be lost in a population despite an organism being self-sacrificial if—and only if—they had the effect of helping a close relative. The relatives concerned must be closely related, since they must carry the sacrificer's genes. In such circumstances, these genes can be selected (that is, survive as a result of this behavior) and spread within families despite the risks to the altruist.[3]

In this way, there could be situations where cooperation (or unselfishness) is an advantage to a group of individuals, even if particu-lar individuals are disadvantaged. Haldane's argument was formalized and extended in 1964 by William D. Hamilton as "inclusive fitness," nowadays often called "kin selection." It was popularized by Edward O. Wilson as a major driver of "sociobiology."

Sociobiological ideas have been extremely important in biology, and have stimulated an immense amount of research. They have also provoked much dissent, particularly as they apply to mammals (especially humankind) because of the implications that behavioral choices are programmed (or determined) by genes. Moreover, human social organization extends far beyond family relationships, and most observers believe that the sociobiological mechanisms so far investigated are insufficient to account fully for the widespread acknowledgment of the Golden Rule ("loving one's neighbors as oneself").

This was another puzzle for Darwin. He wrote in *The Descent of Man*,

> There is no evidence that man was aboriginally endowed with the ennobling belief in the existence of an Omnipotent God. On the contrary there is ample evidence, derived not from hasty travelers, but from men who have long resided with savages, that numerous races have existed, and still exist, who have no idea of one or more gods, and who have no words in their languages to express such an idea. The question is of course wholly distinct from that higher one, whether there exists a Creator and Ruler of the universe, and this has been answered in the affirmative by some of the highest intellects that have ever existed. . . . To do good in return for evil, to love your enemy, is a height of morality to which it may be doubted whether the social instincts would, by themselves, have ever led us. It is necessary that these instincts, together with sympathy, should have been highly cultivated and extended by the aid of reason, instruction, and the love or fear of God, before any such golden rule would ever be thought of and obeyed.

The best solution to this puzzle—of how human morality originated—seems to be to go beyond formal genetics and recognize the reality in higher animals, and most markedly in mankind, of two kinds of heredity. One has led to the evolution of our biological form, our genetic heredity. The other controls our cultural evolution, a very different kind of inheritance. We can call these our *organic* and *superorganic* (or cultural) heredity.

Biological inheritance in humans is exactly the same as that in any other sexually reproducing organism, based on the transmission of genetic information encoded in DNA from one generation to the next by means of the sex cells. Cultural inheritance, on the other hand, is based on transmission of information by a teaching-learning process, which is in principle independent of biological parentage. Culture is transmitted by instruction and learning, by example and imitation, through books, newspapers, radio, television, and motion pictures, through works of art, and by any other means of communication. Every person acquires culture from parents, relatives, and neighbors, and from the whole human environment.

Cultural inheritance enables humans to accomplish what no other organism can do: the transmission of stored experience from generation to generation. Animals can learn from experience, but they do not transmit their "discoveries" (at least not to any large extent) to the following generations. Animals have individual memories, but they do not have a lasting social memory. Humans, on the other hand, have a culture by which they can transmit cumulatively their experiences from generation to generation. Cultural inheritance makes possible cultural evolution—that is, the evolution of knowledge, social structures, ethics, religion, and all other components that make up human culture. Cultural inheritance enables a mode of adaptation to the environment not available to nonhuman organisms.

Cultural adaptation has largely taken over from biological adaptation in humankind both because it is a more rapid mode of adaptation and because it can be directed. A favorable genetic mutation newly arisen in an individual can be transmitted to a sizeable part of a species only through innumerable generations. In contrast, a new scientific discovery or technical achievement can reach the whole of humanity in less than one generation. Moreover, whenever a need arises, culture can directly pursue the appropriate changes to meet the challenge. This differs radically from biological adaptation, which depends on the accidental availability of a favorable mutation (or combination of several mutations), at the time and place where the need arises.

Our ancestry has certainly been shaped by selection. However, the importance of cultural evolution means that we will not fully understand human nature even if we successfully bring together embryology, anatomy, genetics, ecology, and behavioral studies. Certainly such a synthesis would be a worthwhile enterprise, but it would be foolish to expect this to give us all the answers we would like. Nor do we have to go along with Wallace and postulate a guiding "Higher Intelligence."

In all this, we have to convince two sorts of arch-reductionists: those who insist that we are no more than survival machines controlled by selfish genes and those who focus on personal and sexual relationships and group dynamics at the expense of everything else. E. O. Wilson, prophet of a middle road, argues that there are two sorts of people, empiricists and transcendentalists, whose existence should be experimentally verifiable. He believes that we will not progress much until we can identify these different tribes. But he has been pilloried as a modern high priest of naturalism, giving encouragement to such critics of religion like Dawkins and Dennett. Wilson has written: "What was the origin of mind, the essence of mankind? We suggest that a very special form of evolution, the melding of genetic change with cultural history, both created the mind and drove the growth of the brain and the human intellect forward. . . . [We want to] link the research on gene-culture coevolution to other, primarily anatomical studies of human evolution."

On the face of it, nothing is particularly radical in this. Any evolution is the result of interactions between the environment—which includes social forces—and the genetic system. This is no different from conventional ecology. But does it get to the root of humanness? Humans are obviously different from the apes in many ways, but are the differences merely ones of degree or is there a real qualitative difference? This is probably unanswerable from the scientific point of view, but theologically there is a simple solution: to regard the biological species *Homo sapiens*, descended from a primitive simian stock and related to living apes, as having been transformed by God at some time in history into *Homo divinus*, biologically unchanged but spiritually distinct. There is no reason to insist that this event took place at the same time as the emergence of the biological form we call *H. sapiens*. We return to this possibility in the last chapter.

CHAPTER 7
The Most Dangerous Species

IN 1624 JOHN DONNE famously wrote, "No man is an island, entire of itself." Indeed, every species on Earth affects other species—competing for food or space, protecting or being protected, eating or being eaten. These interactions are the subject of ecology; past chapters have described something about them. An important lesson is that no one species is essential; a key implication of island biotas is that communities adjust pragmatically to missing or new species. It is wrong to assume that there is an inevitable collection of species in any particular habitat or that every species exercises an equal effect on its community. There is no automatic balance or equilibrium. Some species can disappear without being noticed; others exert a disproportionate effect relative to their size or numbers. These latter have been called "keystone species." They behave like a keystone in a stone arch: if the keystone is removed, the arch collapses.

The idea of keystone species was proposed by Robert Paine of the University of Washington on the basis of his studies of the shallow waters of the eastern Pacific. A carnivorous starfish, *Pisaster ochraceus*, occurs there. If it disappears (or is removed), two mussel species that are its prey increase in numbers to the extent that they take over the community and crowd out other species, greatly reducing the diversity in the area.

Most species do not play such a major role. There are millions of species; we don't know the exact number (see p. 166). But we do know that one species—*Homo sapiens*—has had a disproportionate effect wherever it occurs. We have changed the Earth in ways no other species could; we have been called the "most dangerous" of all species, using up more of the Earth's resources than any other.

We cannot escape evidence of human influence. Land-use patterns are visible from space; concentrations of carbon dioxide, methane, nitrous oxide, and other gases are increasing in the atmosphere as a consequence of human activities; whales, cod, tuna, tigers, and passenger pigeons are much rarer than they used to be—or absent altogether. Our impacts are everywhere; do we have a role or a responsibility to ease them in any way—reengineering the keystone, as it were? One objection to any such attempt is that nature is so vast and powerful that humans *cannot* alter it in any significant way even if they tried. The paleontologist Steven Jay Gould has written that humans are arrogant to think that, given our small size and frailties, we are capable of affecting the natural world other than trivially. As Gould saw it,

> The views that we live on a fragile planet now subject to permanent derailment and disruption by human intervention [and] that humans must learn to act as stewards for this threatened world ... however well intentioned, are rooted in the old sin of pride and exaggerated self-importance. We are one among millions of species, stewards of nothing. By what argument could we, arising just a geological microsecond ago, become responsible for the affairs of a world 4.5 million years old, teeming with life that has been evolving and diversifying from at least three-quarters of this immense span? Nature does not exist for us, had no idea we were coming, and doesn't give a damn about us.[1]

A similar implication of helplessness comes from the Gaian idea of the Earth as a self-regulating, homeostatic system: if Gaia is at work, any change in the Earth's circumstances would be expected to result in an inevitable readjustment through natural feedback systems.

Notwithstanding, these views are having to change as our knowledge of the real situation adds up. Even Jim Lovelock, the begetter of the Gaia theory, has been forced to accept that the Earth's system is proving unable to cope with the stresses placed upon it, particularly from the vast amounts of greenhouse gases that we have been generating and releasing into the atmosphere. These are causing climatic change, but also knock-on effects: increased atmospheric CO_2 means that more CO_2 is dissolved in the oceans, leading to the sea water becoming more acidic, with implications for the productivity of the plankton on which the marine food chains depend. Lovelock now speaks of "the revenge of Gaia" on those who are assaulting it.

One way of estimating our effect on the Earth is to measure how much of its resources we take. We can calculate—very, very roughly—the Net Primary Production (NPP) of the Earth—that is, how much energy is left after subtracting the respiration of the primary producers (i.e., plants) from the total energy received (mainly from the Sun) and fixed in living systems around the Earth. The NPP involves estimating the maintenance, growth, and reproduction of the primary producers. A crude estimate of NPP is 132 billion tons (see Table 7.1).

The next stage is to put ourselves into this equation. How much of this NPP do we take for ourselves? There are three ways of making such a calculation. A minimum figure is the amount we use directly—for food, fuel, fiber (such as cotton), or timber for building. An intermediate figure takes into account the productivity of land devoted to human activities. But we also ought to consider productivity lost as a result of building and roads, and of misuse leading to desertification or degradation through overgrazing or erosion. This third approach—the highest figure—seems the best estimate our impact. Any calculation is extremely rough, but in total:

- We eat about a million tons of plant growth a year.
- Our domestic animals eat about 2 million tons of plant growth a year.
- We use about 2 billion tons of plant growth each year for building materials and fuel.
- The global area of agricultural crops is 15 million square kilometers (km²), producing 1,700 tons of crops per km², a total production of 26 billion tons per year.
- We have built over land that would have generated 3 billion tons of growth per year.
- Forest converted to grazing land accounts for 3 billion tons of plant production per year.
- We harvest or burn 1,500 tons of temperate and boreal trees per km² per year, a total of three-quarters of a billion tons.
- Tree plantations make up 3 billion tons of biomass used per year.
- We clear 160,000 km² of tropical forest each year, but fires and selective logging damage several times more than this yearly.
- At least 5 million km² of tropical forest have been converted to pasture or otherwise changed.
- Salt accumulation from irrigation destroys the productivity of 15,000 km² of cropland per year and reduces the productivity of 450,000 km².
- Erosion and other agricultural practices in arid regions destroy about 20,000 km² each year and reduce the productivity of another 2 million km².
- About 6 million km² of pasture lands are burned each year.
- Overgrazing has damaged productivity over 35 million km² drier lands.

Adding all this up, our activities apparently account for the consumption (or loss of) 60 billion tons of actual or potential primary productivity per year. We harvest 26 billion tons directly from agricultural crops and 14 billion tons from forests, but also lose

TABLE 7.1 Global Primary Production

Ecosystem type	Mean net primary productivity (g/m²/yr)	Mean biomass (kg/m²)
Continental		
Tropical rain forest	2,200	45.0
Tropical seasonal forest	1,600	35.0
Temperate evergreen forest	1,300	35.0
Temperate deciduous forest	1,200	30.0
Boreal forest	800	20.0
Woodland and shrub	700	6.0
Savannah	900	4.0
Temperate grassland	600	1.60
Tundra and alpine	140	0.60
Desert and semidesert scrub	90	0.70
Extreme desert, rock, sand, ice	3	0.02
Cultivated land	650	1.00
Swamp and marsh	2,000	15.00
Lakes and streams	250	0.02
Mean continental (adjusted for area of each habitat type)	**773**	**12.3**
Marine		
Open ocean	125	0.003
Upwelling zones	500	0.02
Continental shelf	360	0.01
Algal beds and reefs	2,500	2.0
Estuaries	1,500	1.0
Mean marine (adjusted for area)	**152**	**0.01**
Grand total production	**333**	**3.6**

Following Peter Vitousek et al., "Human Appropriation of the Products of Photosynthesis," *Bioscience* 36 (1986): 368–73.

the use of three billion tons because of land loss through buildings and another 17 billion tons from grazing. In other words, we are using 45 percent of the Earth's 132 billion tons NPP. A similar, albeit less precise, calculation for fisheries suggests that we are taking 35 percent of the productivity of the continental shelves—the most productive areas of the seas. We also use 60 percent of available freshwater.

When we add land and sea, we cannot avoid the conclusion that we have a very major impact on the Earth. This is where we are at the moment. But we also have to face the fact that the human population will grow by another 50 percent within the next century, never mind certain but unquantified changes in climate. It has been truly said that we are threatened by a perfect storm.

Running Out of World

Faced with everything we are taking from the Earth, we can be said to be running out of world. This has been happening on a regional scale throughout history; the difference now is that it is happening on a global scale. China alone has had around two thousand famines in the last two thousand years; overpopulation and land scarcity have led to successive mass population movements; the Beaker Folk, Teutons, Vikings, and New World colonizers all spilled from the western seaboard of Europe. Mismanagement has often produced disastrous consequences for human: the early Polynesian population of New Zealand depended on the large flightless Moa for food, but managed to drive it to extinction within six hundred years; introduction of rabbits to Australia, mongooses to Hawaii, and gray squirrels to Britain have all caused major ecological problems. Overextension of irrigation was a major factor in the collapse of the ancient Babylonian empire. Sicily was once the granary of Italy, but less and less corn is grown there as the soil deteriorates under excessive cultivation and goat browsing. The ecological implosion of Easter Island is well documented (see Figure 7.1).

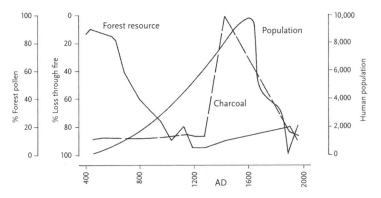

FIGURE 7.1. Easter Island lies in the central Pacific. It has suffered a series of ecological assaults through overexploitation of resources: deforestation (shown by the pollen record), resulting in lack of wood for boats and soil erosion; overpopulation, exacerbating soil erosion and leading to territorial disputes; predation by rats and the elimination of many birds (courtesy John Flenley; based on Paul Bahn and John Flenley, *Easter Island, Earth Island* [New York: Thames & Hudson, 1992]).

Despite all this, we continue to behave as if we can carry on without change, probably because our perception of space is so often ambiguous. On one hand, we have become familiar with images of the Earth floating in space, emphasizing our finiteness and the reality of our limited resources, particularly when supplemented by instant television pictures of degraded landscapes, reinforcing our sense of frailty and insignificance. On the other hand, common observation—including the large distances we cover in long-distance plane flights—shows us a world very big compared to our physical size, encouraging us to believe that our impact will be so puny that it will have no significant effect on the world as a whole. We know the facts but dismiss their implications as scaremongering; in the affluent North we too often close our eyes to poverty and the effects of climate change in Africa or on small island states.

In the face of this ambiguity, we cannot even agree on the most correct analysis of our environmental problems. In the 1960s the debate was between proponents of "zero population growth" and those who saw salvation coming from a more efficient use of resources. These kinds of debate should continue, but the major obstacle to progress is more all-embracing; it is the failure of ecological literacy. We need to take global environmental issues seriously and face our keystone effect as a species. We have been very slow to do this.

MODERN-DAY WARNINGS

Where are we now? In 1964 Rachel Carson's book *Silent Spring* was a wakeup call to insidious environmental damage. The book turned out to have a very large influence in creating environmental awareness, particularly in North America. It also marked the beginning of actions by governments, groups of scientists, and religious institutions. Key events since then are as follows:

1972   *Limits to Growth* from M.I.T.; U.N. Conference on the Human Environment, Stockholm; establishment of U.N. Environment Programme

1980   *World Conservation Strategy*

1982   World Charter for Nature accepted by the United Nations

1983   World Council of Churches establishes Programme for Justice, Peace and the Integrity of Creation (JPIC)

1987   *Our Common Future* (World Commission on Environment and Development)

1988   U.N. Intergovernmental Panel on Climate Change set up

1991   *Caring for the Earth* (revised *World Conservation Strategy*)

1992   Earth Summit at Rio de Janeiro (U.N. Conference on Environment and Development), agreeing to Conventions on Climate Change (UNFCCC) and Biological Diversity (CBD), Agenda 21, Rio Declaration

1995   Draft of International Covenant on Environment and Development (ICED)

1997   First draft of Earth Charter

2000   Millennium Development Goals

2005   Millennium Ecosystem Assessment (MA) published

2010   Intergovernmental Platform on Biodiversity and Ecosystem Services (IPBES)

From all these documents and conferences, a few general principles have emerged. As we will see, the principles reappear across these efforts. They can be seen in particular in the Earth Charter and the International Covenant on Environment and Development (see p. 179).

But what should perhaps concern us is that environmental concerns are not new; some of them go back centuries. As long ago as 1758, a Danish pastor, Otto Lütken, wrote:

> Since the circumference of the globe is given and does not expand with the increased number of its inhabitants, and as travel to other planets thought to be habitable has not yet been invented; since the earth's fertility cannot be extended beyond a given point and since human nature will presumably remain unchanged, so that a given number will hereafter require the same quantity of the fruits of the earth for their support as now, and as their rations cannot be arbitrarily reduced, it follows that the proposition "that the world's inhabitants will be happier, the greater their number" cannot be maintained, for as soon as the number exceeds that which our planet with all its wealth of land and water can support, they must necessarily starve one another out, not to mention other necessarily attendant inconveniences, to wit, a lack of other

comforts of life, wool, flax, timber, fuel and so on. But the wise Creator who commanded men in the beginning to be fruitful and multiply, did not intend, since He set limits to their habitation and sustenance, that multiplication should continue without limit.

The notion of limits surfaced—more accurately, resurfaced—after the Second World War, when UNESCO sponsored a global gathering of scientists. This group went on to form the International Union for Conservation of Nature (IUCN) based in Gland, Switzerland, which since 1948 has played a pivotal role in many of the events that came later.

Lütken's warning has been all too graphically illustrated by Jared Diamond in a book, *Collapse*, published in 2005. He identified five causes for the collapse of cultures: (1) environmental damage, (2) climate change, (3) hostile neighbors, (4) trade partners, and (5) response to symptoms of community or environmental stress. He noted that the first four of these vary in importance in different situations, but the last is always significant (and relates to our use of resources).

Silent Spring was a major wakeup call to gratuitously ignoring natural processes. Eight years later (in 1972), a computer simulation carried out at the Massachusetts Institute of Technology and published as *The Limits to Growth* showed that the economic and industrial systems of developed countries would collapse about the year 2100 unless birth and death rates equalized and capital investment matched capital depreciation. Economists disliked and savaged the *Limits* study on the grounds that it ignored market forces and technological developments. Notwithstanding, it resonated widely in drawing attention to the fact that a finite system has inevitable boundaries, even if we cannot agree what these are or when we shall reach them. (Follow-up analyses twenty and thirty years after the original report using better data and refined programs merely confirmed the initial conclusions.)

The *Limits* study highlighted the danger of ignoring our impact; our survival depends ultimately on our use of resources and whether or not they are renewable. It was published to coincide with the first major international conference on the environment, the U.N. Conference on the Human Environment, held the same year in Stockholm. The Stockholm Conference is generally credited with introducing the concept of sustainability into general discourse. It declared, "A point has been reached in history when we must shape our actions throughout the world with a more prudent care for their environmental consequences. Through ignorance or indifference we can do massive and irreversible harm to the earthly environment on which our life and well-being depend." A basic concept was "development without destruction."

Stockholm stimulated a tremendous amount of concern and support for development in the developing world, but also produced a division between activists for "development" and "environmental protection." Environmental care was regarded as a much lower priority than attacking poverty; indeed, environmental care was commonly regarded as a hindrance to development. In 1980 a World Conservation Strategy was produced to counter these assumptions.

World Conservation Strategy

In 1980 the United Nations commissioned a forum of experts, led by the IUCN, to draw up a plan for global conservation. The result was the *World Conservation Strategy* (*WCS*). It set out three aims: (1) to maintain essential ecological processes and life-support systems, (2) to preserve genetic diversity, and (3) to ensure the sustainable utilization of species and ecosystems. The *Strategy* focused firmly on people:

> Humanity's relationship with the biosphere (the thin covering of the planet that contains and sustains life)

will continue to deteriorate until a new international economic order is achieved, a new environmental ethic adopted, human populations stabilize, and sustainable modes of development become the rule rather than the exception.

Sustainability became part of the international vocabulary. Its implications were taken up by a World Commission on Environment and Development, chaired by Gro Harlem Brundtland, whose 1987 report, *Our Common Future* (commonly referred to as the Brundtland Report), emphasized the need to recognize ecological as well as economic interdependence among nations.

The Brundtland Report is well known for its definition of sustainable development: *that which meets the needs of the present without compromising the ability of future generations to meet their own needs.* Although oft-quoted, this definition has been heavily criticized as ambiguous and open to contradictory interpretations. Importantly, it ignores the "limits to growth" constraint, implying that "nature" has the capacity to meet all human needs if social and technological deficiencies are sorted out. This belief in unlimited resources still persists in some places.

A revised *WCS* (published as *Caring for the Earth,* 1991) redefined "sustainable development" in a more ecologically aware way as *improving the quality of human life while living within the carrying capacity of supporting ecosystems. Caring for the Earth* also pointed out the confusion around the uses of "sustainable" as an adjective:

> "Sustainable development," "sustainable growth," and "sustainable use" have been used interchangeably as if their meanings were the same. They are not. "Sustainable growth" is a contradiction in terms: nothing physical can grow indefinitely. "Sustainable use" is applicable only to renewable resources: it means using them at rates within their capacity for renewal. . . . A "sustainable economy"

is the product of sustainable development. It maintains its natural resource base. It can continue to develop by adapting, and through improvements in knowledge, organization, technical efficiency and wisdom.

Caring for the Earth also called for "a world ethic for living sustainably." A significant criticism of the original *Strategy* had been that it fell into a fallacy inherited from the Enlightenment by failing to emphasize that responsible behavior toward the environment did not automatically follow from recognizing environmental facts. *Caring for the Earth* detailed the ethic missing from the original *World Conservation Strategy*; it included an aim that everyone should

> share fairly the benefits and costs of resource use, among different communities and interest groups, among regions that are poor and those that are affluent, and between present and future generations. Each generation should leave to the future a world that is at least as diverse and productive as the one it inherited. Development of one society or generation should not limit the opportunities of other societies or generations.

The results of this call for an ethic and its possible content are set out in a later section. Increasingly it has been realized that scientific understanding of the environment is insufficient by itself to bring about responsible environmental attitudes of care and maintenance. Environmental care requires as full a knowledge of facts and processes as possible, but also an ability to balance personal with community interest plus an assessment of the value of the environment to all involved.

Caring for the Earth set out the problems and possible ways forward for the United Nations Conference on Environment and Development, commonly called the Earth Summit, convened in Rio de Janeiro in 1992. One hundred seventy-two nations participated, 108 of them represented by their heads of state or government. The conference agreed on a "Rio Declaration" (listing twenty-seven principles intended as a guide toward sustainable development), Agenda 21 (setting out an "action plan" for the twenty-first century), and two conventions—a Framework Convention on Climate Change (FCCC) and a Convention on Biological Diversity (CBD). The Convention on Climate Change is the better known of the two because of its subsequent history and continuing controversy. It declared that climate change is a serious problem, that action cannot wait upon the resolution of scientific uncertainties, that the developed countries should take a lead in action, and that the developed countries should compensate developing countries for any additional costs incurred on taking measures under the convention.

PROTECTING BIODIVERSITY

The Convention on Biological Diversity aimed to preserve the biological diversity of the planet through the protection of species and ecosystems and to establish conditions for the associated uses of biological sources and technology. It affirmed that states have "sovereign rights" over biological resources in their territory, the fruits of which should be shared in a "fair and equitable" way on "mutually agreed terms." One hundred and ninety-three states have signed up to it. The only nation that has so far positively refused to do so is the United States of America, on the grounds that it would impose a restrictive framework for prospecting on organisms that could be useful for the development of drugs or other commercially valuable compounds, including their potential use for genetic modifications, and that it would not protect intellectual property rights.

It soon became clear that data were needed to support the CBD. There were no organizations in the biological sphere like the Inter-Governmental Panel on Climate Change, set up in 1988 by the

United Nations Environment Programme and the World Meteorological Organization. There were even no accurate data about the extent of biodiversity. Millions of species of animals and plants live on Earth. Over the centuries, the combined efforts of naturalists and museum specialists have given names to about 1.8 million. Everyone agrees, however, that this is a mere fraction of the total (see Table 7.2).

We can be pretty certain about the total number of mammal and bird species, although a few more are recognized every year. Over the past eighty years, for example, an average of one mammal species a year has been described, most of them tropical bats, rodents, shrews, or small marsupials. Of the roughly 10,000 species of birds listed, about 140 were identified over this same period, mostly small, brown tropical birds. For obvious reasons, invertebrates—with the exception of butterflies—are a more difficult task to collect and list.

Twenty or thirty years ago, it was thought that the total number of species might be three times or so the already catalogued ones. This calculation was based on two observations. First, among well-studied groups like birds and mammals, there are roughly two tropical species for every temperate or boreal one; second, around two-thirds of all named species are found outside the tropics. In other words, if the ratio of numbers of tropical to non-tropical species is the same for insects as for birds and mammals, we may expect something like two yet-unnamed species of tropical species for every one currently named.

Even so, the guess of 3 to 5 million that results from this calculation seemed a low estimate, because when we get down to some of the invertebrate and "lower plant" groups, never mind microorganisms and parasites, there is no doubt that the majority of forms are still awaiting discovery and naming.

Indeed this guesstimate has been challenged by studies of insects living in the canopies of tropical trees. Many of these insects are confined to feeding on particular tree species. Remarkably, there

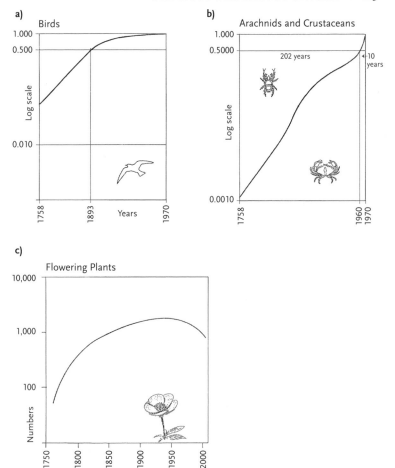

FIGURE 7.2. Declining number of new species described since the publication of the 10th edition of Linnaeus's *Systema Naturae* in 1758. a) Rate of description of new bird species—half the species known in 1970 had been recorded by 1843. b) Rate of discovery of arachnids and crustaceans—half the 1970 total were described after 1960 (from Robert M. May, *Philosophical Transactions of the Royal Society of London* 330 [1990]: 293–304). c) Number of new species of flowering plants (based on data in L. N. Joppa, D. L. Roberts, and S. L. Pimm, *Proceedings of the Royal Society of London*, Series B, 466 [2010]: 298–304). Figures used by permission of the Royal Society of London.

may be as many as 600 insect species feeding on any one tree species (including insects feeding on other insects). Assuming there are approximately 50,000 tropical tree species, this leads to an estimate of 30 million tropical insect species, or perhaps 50 million species in all. This is a maximum figure and may be much too high. The truth is that we are woefully ignorant of the true figure.

TABLE 7.2 Numbers of Species

	Described species	Possible numbers	
		Highest figure	Working figure
Viruses	5,000	500,000+	500,000
Bacteria	4,000	3,000,000+	400,000
Fungi	70,000	1,500,000+	1,000,000
Protozoans	40,000	100,000+	200,000
Algae	40,000	10,000,000+	200,000
Plants	250,000	500,000+	300,000
Vertebrates	45,000	50,000+	50,000
Nematodes	15,000	1,000,000+	500,000
Mollusks	70,000	180,000+	200,000
Crustaceans	40,000	150,000+	150,000
Arachnids	75,000	1,000,000+	750,000
Insects	950,000	100,000,000+	8,000,000

From World Conservation Monitoring Center, *Global Biodiversity* (London: Chapman & Hall, 1992)

This legacy of uncertainty about species has been handed down year after year. Through the 1990s, advances were being made in ecological sciences and resource economics, but their findings were not reflected in responding to the Convention on Biological Diversity or in policy discussions concerning ecosystems. Moreover, the lack of an international forum for collecting and analyzing biodiversity data meant that governments were able to evade the topic. (In contrast, the Intergovernmental Panel on Climate Change provided regularly updated data for debates and regulations relating to climate change.)

This gap led to a call by the United Nations Environment Programme and the World Bank for "a more integrative assessment process for selected scientific issues, a process that can highlight the linkages between questions relevant to climate, biodiversity, desertification, and forest issues." This, in turn, stimulated a Millennium Ecosystem Assessment (MA) by more than thirteen hundred ecologists throughout the world. Over a four-year period, this project assessed the consequences of ecosystem change for human well-being. This made it possible to plan a basis for action to enhance the conservation and sustainable use of those systems. The MA report appeared in 2005 and provided a state-of-the-art scientific appraisal of the condition and trends in the world's ecosystems and the services they provide (such as clean water, food, forest products, flood control, and natural resources) and the options to restore, conserve, or enhance the sustainable use of ecosystems.

The bottom line of the MA findings was that human actions are depleting Earth's natural capital, putting a strain on the environment to the extent that the ability of the planet's ecosystems to sustain future generations can no longer be taken for granted. At the same time, the assessment argued that with appropriate actions it should be possible to reverse the degradation of many ecosystem services over the next fifty years. To do this, however, governments and citizens would have to change their behavior significantly and the necessary changes in policy and practice were nowhere underway.

The data the MA gathered were important in three ways:

First, it represented the consensus view of the largest body of social and natural scientists ever assembled. They identified where broad agreements existed on findings and where information was insufficient to reach firm conclusions for action. The existence of such a broad consensus view is an important contribution to decision-making.

Second, the focus on ecosystem services and their link to human well-being and development needs was new. By examining the environment through the framework of ecosystem services, it became much easier to identify how changes in ecosystems influence human well-being and hence to provide information in a form that decision-makers can weigh alongside other social and economic information.

Third, the assessment identified a number of "emergent" findings, conclusions that could only be reached when a large body of existing information is examined together. Four of these stand out:

1. *The balance sheet.* Although individual ecosystem services had been assessed previously, the recognition that 60 percent of the ecosystem services examined by the MA are being degraded provided the first comprehensive audit of the status of the Earth's natural capital.

2. *Nonlinear changes.* Nonlinear (accelerating or abrupt) changes had been identified previously in a number of ecosystem studies. The MA was the first to conclude that ecosystem changes are increasing the likelihood of such nonlinear changes and to warn of the important consequences of this for human well-being. Examples of changes of this nature include the appearance of new diseases, abrupt alterations in water quality, the creation of dead zones in coastal waters, collapse of fisheries, and shifts in regional climate.

3. *Drylands.* Because the assessment focused on the linkages between ecosystems and human well-being, a somewhat different set of priorities emerged from it than from earlier studies. While the MA confirmed that major problems exist with tropical forests and coral reefs, from the standpoint of linkages between ecosystems and people, the most significant challenge was seen to involve dryland ecosystems. These ecosystems are particularly

fragile, and they are also the places where the human population is growing most rapidly, biological productivity is least, and poverty most acute.

4. *Nutrient loading.* The MA confirmed the emphasis that decision-makers were already giving to important drivers of ecosystem change such as climate change and habitat loss. But the MA also found that excessive nutrient loading of ecosystems is one of the major drivers of change at present and one that will grow significantly worse in the coming decades unless action is taken. The issue of excessive nutrient loading, although well studied, was seen as not yet receiving significant policy attention in many countries or internationally (see Figure 7.3).

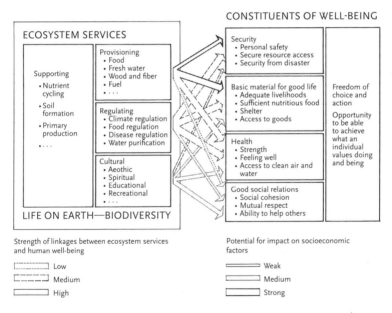

FIGURE 7.3. Consequences of ecosystem change for human well-being (from *Millennium Ecosystem Assessment Synthesis Report* [Washington, DC: Island Press, 2005]).

Around the time that the Millennium Ecosystem Assessment was being set up, in 2000 the world's leaders made a *Millennium*

Declaration, comprising eight Millennium Development Goals to be achieved by 2015. These have served as an idealistic target for politicians and policy makers. The seventh goal is particularly relevant here. It is to "ensure environmental sustainability" and calls for (1) new government policies to integrate sustainable development into national policies and programs, (2) a reduction in the rate of biodiversity loss, (3) increasing access to safe drinking water and basic sanitation, and (4) improving the lives of at least 100 million slum dwellers by 2020.[2]

By identifying specific targets in this way, the Millennium Development Goals drew attention to the impacts that humans are having—and have had—on the environment. The sad likelihood that achievement of the goals is improbable, as was clear from the Millennium Ecosystem Assessment to which we now return.

The MA's findings were:

▶ Over the past fifty years, humans have changed ecosystems more rapidly and extensively than in any comparable period of time in human history, largely to meet rapidly growing demands for food, freshwater, timber, fiber, and fuel. For example, in the last few decades 20 percent of the world's coral reefs have been lost and another 20 percent degraded; 35 percent of the area covered by mangroves has been lost; withdrawal of water from rivers and lakes has doubled in the last fifty years; more than half of all synthetic nitrogen has been applied as fertilizer since 1985.

▶ Changes made to ecosystems have contributed to substantial net gains in human well-being and economic development, but these gains have been achieved at the cost of the degradation of many ecosystem services, increased risks of nonlinear changes, and the exacerbation of poverty for some groups of people. Food production has more than doubled since 1960 and its price has dropped, but 25 percent of commercial fish stocks are being overharvested (the marine fish catch has been declining since the late 1980s).

▸ The degradation of ecosystem services could grow significantly worse during the first half of this century, presenting a barrier to achieving the Millennium Development Goals. Over 15 percent (and possibly as much as 35 percent) of the water used for irrigation is greater than the rate of replenishment. In many areas, pest control by natural enemies has been replaced by artificial pesticides, reducing the capacity of natural systems to provide pest control, and a global decline is occurring in the abundance of pollinators. Once a threshold of nutrient loading is exceeded, changes in freshwater and coastal ecosystems can be abrupt and extensive, creating harmful algal blooms, sometimes leading to the formation of oxygen-depleted zones, killing all animal life. The loss of wetlands and mangroves as well as deforestation has reduced the capacity for natural buffering against extreme events. The economic value of managed ecosystems is almost always higher than new or converted ones (e.g., traditional forest use over timber harvesting; protection from mangroves over shrimp farming), although local gains may complicate the equation (e.g., tourism developments). Deforestation generally leads to decreased rainfall. Since forest existence depends on rainfall, forest loss can result in a positive feedback and exacerbate the situation—accelerating the rate of decline in rainfall, and increasing the loss of forest cover. Desertification affects the livelihoods of millions of people, including a large portion of the poor in drylands. The collapse of the Newfoundland cod fishery in 1992 cost an estimated $2 billion in income support and retraining for the fishermen thrown out of work. The frequency and impacts of floods and fires has increased significantly in the last fifty years, in part at least to environmental changes. Half the urban population in Africa, Asia, Latin America, and the Caribbean suffers from one or more diseases associated with inadequate water and sanitation.

▸ The challenge is to reverse the degradation of ecosystems

while meeting increasing demands for their services. This challenge could be met partially under some scenarios suggested by the MA, but they all involve significant changes in policies not currently under way. For example, a further 10 to 20 percent of grassland and forest is likely to be converted to agricultural use by 2050, and overfishing continues virtually everywhere. Invasive alien species are causing increasing problems. At the same time, the demand for food crops is projected to grow by 70 to 85 percent by 2050, and water withdrawal by 30 to 85 percent.

The Millennium Ecosystem Assessment focused on the fact that nearly all the barriers to solving the ecological problems it identified are social or political. In other words, their solutions do not have any inherent scientific or natural barriers. The obstacles to their implementation are poor institutional and governance arrangements in many countries, including the presence of corruption and weak systems of regulation and accountability. But they include also market failures and misalignment of economic incentives. Many groups dependent on ecosystem services or harmed by their degradation lack political and economic power. There is underinvestment in the development and diffusion of appropriate technologies. Overall, there is insufficient knowledge (and poor use of existing knowledge) about ecosystem services and responses that could enhance benefits from these services while conserving resources.

THE THEOLOGICAL RESPONSE

The Ecosystem Assessment and the targets in the Millennium Development Goals are part—but only a part—of an increasing global perception that we are extracting an unsustainable amount of natural capital. As already noted, an alarm was sounded by the Stockholm Conference in 1972 and reinforced by the World Conservation Strategy in 1980. It needed a moral response.

Two theologians took up the challenge. At the 1983 Assembly

of the World Council of Churches (WCC), the German theologian Jürgen Moltmann argued that the traditional Christian call for "peace with justice" was futile unless it took place within a *whole* creation, a creation with "integrity." His advocacy led to the replacement of the need for a "Just, Participatory and Sustainable Society," which had dominated WCC policy in the 1970s, by a more inclusive "Justice, Peace and the Integrity of Creation" (JPIC) program.

This change was welcomed by developing countries, who had (rightly or wrongly) associated "sustainability" as a continuance of colonial injustice; for them, JPIC implied the rejection of a global hegemony in favor of regional associations A key was that environmental concern should be seen as integral to "justice" and "peace". The concept of the "integrity of creation" sought to convey the dependence of creation on its Creator and the worth and dignity of creation in its own right.

The JPIC process progressed with different emphases and hopes in different parts of the world. It culminated in a global consultation in Seoul in 1990, which revealed more discord than harmony. A significant minority of participants saw meaning in creation through mysticism or traditional religions. Such theological pluralism was unacceptable to most of those present, reducing the impact that JPIC might have had.

The second theologian with an environmental agenda was another German, Hans Küng. His initiative was specifically multi-faith, emanating from the Parliament of the World's Religions meeting in Chicago in 1993. Küng proclaimed,

> We are convinced of the fundamental unity of the human family on Earth. We recall the 1948 Universal Declaration of Human Rights of the United Nations. What is formally proclaimed on the level of rights we wish to confirm and deepen from the perspective of an ethic. . . . Earth cannot be changed for the better unless the consciousness of individuals is changed. We pledge to work

for such transformation in individual and collective consciousness, for the awakening of our spiritual powers through reflection, meditation, prayer, or positive thinking, for a conversion of the heart. Therefore we commit ourselves to a common global ethic, to better understanding, as well as socially-beneficial, peace-fostering, and Earth-friendly ways of life. . . . Limitless exploitation of the natural foundations of life, ruthless destruction of the biosphere, and militarization of the cosmos are all outrages. As human beings we have a special responsibility—especially with a view to future generations—for earth and cosmos, for air, water, and soil. We are *all intertwined together* in this cosmos and are all dependent on each other. . . . Therefore the dominance of humanity over nature and the cosmos must not be encouraged.

THE EARTH CHARTER

A similar sentiment, but one arrived at by a very different and non-religious route, was the Earth Charter Initiative. One of the original aims of the Earth Summit in Rio in 1992 was to produce an Earth Charter as "a short, uplifting, inspirational, and timeless expression of a bold new global ethic," a statement containing "the basic principles for the conduct of nations and peoples with respect to environment and development to ensure the future viability and integrity of the Earth as a hospitable home for human and other forms of life." It was an aspiration that disappeared in pre-conference wrangling. The charter was changed into a less focused set of statements, issued and agreed to as the "Rio Declaration."

However, there was still a feeling that there are indeed basic principles that could be expressed as an Earth Charter. As a result, Maurice Strong (who had been secretary-general of both the Stockholm and Rio Conferences) and ex-Russian president, Mikhail Gorbachev, independently launched organizations to pursue this aim. Their organizations came together in 1994, and a draft Earth Charter was published three years later; a definitive version appeared in 2000.

The Earth Charter is an attempt "to promote the transition to sustainable ways of living and a global society founded on a shared ethical framework that includes respect and care for the community of life, ecological integrity, universal human rights, respect for diversity, economic justice, democracy, and a culture of peace." It sets out sixteen principles in four groups:

I. *Respect and care for the community of life*
   1. Respect Earth and life in all its diversity.
   2. Care for the community of life with understanding, compassion, and love.
   3. Build democratic societies that are just, participatory, sustainable, and peaceful.
   4. Secure Earth's bounty and beauty for present and future generations.

II. *Ecological integrity*
   5. Protect and restore the integrity of Earth's ecological systems, with special concern for biological diversity and the natural processes that sustain life.
   6. Prevent harm as the best method of environmental protection, and when knowledge is limited, apply a precautionary approach.
   7. Adopt patterns of production, consumption, and reproduction that safeguard Earth's regenerative capacities, human rights, and community well-being.
   8. Advance the study of ecological sustainability, and promote the open exchange and wide application of the knowledge required.

III. *Social and economic justice*
   9. Make the eradication of poverty an ethical, social, and environmental imperative.

10. Ensure that economic activities and institutions at all levels promote human development in an equitable and sustainable manner.

11. Identify gender equality and equity as prerequisites to sustainable development and ensure universal access to education, health care, and economic opportunity.

12. Uphold the right of all, without discrimination, to a natural and social environment supportive of human dignity, bodily health, and spiritual well-being, with special attention to the rights of indigenous peoples and minorities.

IV. *Democracy, nonviolence, and peace*

13. Strengthen democratic institutions at all levels, and provide transparency and accountability in governance, inclusive participation in decision making, and access to justice.

14. Integrate into formal education and lifelong learning the knowledge, values, and skills needed for a sustainable way of life.

15. Treat all living beings with respect and consideration.

16. Promote a culture of tolerance, nonviolence, and peace.

As a bald statement, the charter explicitly endorses sensitivity and concern for human and natural life. What is not apparent from such a seemingly arbitrary set of statements is their dependence on earlier gropings toward a global ethic. Beginning from the awareness that we are running out of world and its quantification in the computer simulations of the *Limits to Growth* studies, and with the concerns about poverty highlighted at the Stockholm Conference and ideals expressed in a Word Charter for Nature agreed by the United Nations in 1982, there was a progression made clear in the World Conservation Strategy, with its clear demonstration that sustainable development is impossible without reliable environmental care. But we have seen, this train of connections was not sufficient. Its implementation depended on personal and political commitment; reliable environmental care is only attainable through responsible management on the part of individuals and organizations, including states. We may be apart from nature, but we are also a part of it.

There was a conspicuous weakness in all these analyses. There was a lack of motivation for action; there was no ethical imperative. This was recognized by the International Union for the Conservation of Nature (IUCN) (the main author of the *World Conservation Strategy*), which set up an Ethics Working Group to influence the content of *Caring for the Earth* (the 1991 updated version of the *WCS*). Meanwhile, the European Commission in 1989 hosted a Conference on Environmental Ethics for the Economic Summit Nations (G7).

Opening the conference, Jacques Delors, then president of the commission, pointed out that the French rationalist philosopher René Descartes' call for humankind to "render ourselves masters and possessors of nature" had led to a search for control that was characteristic of industrial societies everywhere, not only in the West or limited to the inheritors of the Judeo-Christianity of Descartes. Delors concluded, "Despite differing traditions, the right to use or exploit nature seems to have found in industrial countries the same favour, the same freedom to develop, the same economic justification."

Faced with this, Delors argued that we must modify our behavior toward the environment since none of our problems can be approached separately:

> They highlight [our] dependence on the environment, hitherto ill-perceived. They underscore the sudden fragility of man's relationship with nature, which has traditionally been one of mastery based on use and exploitation. It is in the broadest sense the very conditions of humanity which current problems compel us to rethink and rebuild, insofar as continuing our traditional modes of life on earth would lead us to ever-increasing damage and before very long threaten to destroy us.

At almost the same time, the secretary-general of the United Nations, Javez Perez de Cuellar, called in his 1990 report for action on the environment: "The Charter of the United Nations governs relations between States. The Universal Declaration of Human Rights pertains to relationships between the State and the individual. The time has come to devise a covenant regulating relations between humankind and nature."

ETHICS AND THE ICED

The International Environmental Law Commission responded to the secretary-general with an International Covenant on Environment and Development (ICED), intended to codify into hard law the large amounts of soft law on the environment contained in a range of agreements and treaties such as the Stockholm Conference, the World Charter for Nature, the Law of the Sea, and the Rio Declaration. The draft covenant has been through various revisions, most recently in 2004. It still awaits formal adoption by the international community.

The ICED consists of a preamble and seventy-two articles, arranged in eleven sections. Part I states the object of the charter ("To achieve environmental conservation and sustainable development by establishing integrated rights and obligations"), while part II contains "the most widely accepted and established concepts and principles of international environmental law," ideas that repeatedly appear in international discussions and agreements. The most important of these are the declaration produced at the end of the Stockholm Conference, the World Charter for Nature, and the Rio Declaration, but these are only the tip of an iceberg. At least seventy statements or agreements listing an ethic or movements toward an ethic have been produced. Intriguingly, the principles set out in the early parts of ICED are very similar to the Principles that form the core of the Earth Charter.

For our purposes here it is possible to reduce the essential ideas to ten premises, all of which are found in previously published documents (Table 7.3). This commonalty suggests that they really do represent truly fundamental principles rather than convenient generalizations.

TABLE 7.3 Ten Premises for Sustainable Living

These principles are based on convergence between the International Covenant for Environment and Development (currently "deposited" with the United Nations) and a "Code of Environmental Practice" accepted by the G7 Heads of State in 1991. In the following, "S" refers to the Stockholm Declaration that came from the UN Conference on the Human Environment in Stockholm in 1972, "C" to the World Charter for Nature produced by the UN in 1982, and "R" to the Rio Declaration passed by the UN Conference on Environment and Development (the Earth Summit) at Rio in 1992.

1. Environmental conservation and sustainable development are essential for human health and well-being on a planet with finite resources and carrying capacity (S13; C2; R1, 4).

2. Nature as a whole warrants respect; every form of life is unique and is to be safeguarded independently of its worth to humanity (C Preamble, 1).

3. The global environment both within and beyond the limits of national jurisdictions is a common concern of humanity, held in trust for future generations by the present generation. All persons have a duty to protect and conserve the environment; each generation has a responsibility to recognize limits to its freedom of action and to act with appropriate restraint, so that future generations inherit a world that meets their needs (S2, 4; C3; R4).

4. To achieve sustainable development, environmental protection and management must be an integral part of all development efforts. States have, in accordance with the Charter of the United Nations and the principles of international law, not only the sovereign right to their own resources, but the responsibility

a. to protect and preserve the environment within their jurisdiction or control;

b. to ensure that activities within their jurisdiction or control do not cause serious damage to the environment of other states or to areas beyond the limits of national jurisdiction;

c. to work with and collaborate in good faith with other states and competent governmental and nongovernmental organizations in the implement of the covenant;

d. to minimize waste in the use of natural resources and ensure that renewable natural resources are maintained sustainably, and to develop and adopt the most efficient and environmentally safe technologies for the harnessing and use of energy (S7, 14, 21; C4, 7; R4, 12).

5. All states and all people shall cooperate in promoting health, social well-being, and environmental quality by striving to eradicate poverty; this is an indispensable requirement for both sustainable development and distributive justice, and can be achieved only by eliminating unsustainable patterns of production and consumption and by promoting appropriate demographic policies (R5, 7, 8).

6. States have a responsibility to anticipate, prevent, and minimize significant adverse effects of human activities on the environment; lack of full scientific certainty must not be used as a reason to postpone action to avoid potential harm to the environment (C11; R15, 18, 19).

7. States shall take all necessary measures to ensure that the full costs of prevention or compensation for environmental damage, as well as the costs of restoration of the environment, are borne by the person or organization whose activities give rise to such damage or the threat thereof, unless such obligations are otherwise allocated by national or international law. States have the right to be protected against or compensated for significant environmental harm caused by activities outside their own jurisdiction (S22; C12; R13).

8. States shall require environmental impact assessments for all proposed activities likely to have a significant environmental effect and shall include the full social and environmental costs of all environmental impacts within the calculation of those effects (C11; R17).

9. States shall establish and maintain a legal, administrative, research, and monitoring framework for environmental conservation, giving

full and equal consideration to environmental, economic, social, and cultural factors. In particular, states shall

a. regularly review their policies on the integration of planning and development activities and publish their findings;

b. develop or improve mechanisms to facilitate the involvement of concerned individuals, groups, organizations, indigenous peoples, and local communities in environmental decision making at all levels, and provide effective access to judicial and administrative proceedings affecting the environment;

c. make clear the full social and economic costs of using natural resources and ensure the equitable distribution of income generated (S18, 20; R9, 10, 22).

10. Justice, peace, development, and environmental protection and management are interdependent and indivisible, and vital to the integrity of creation (S1; R25). States have a responsibility to work toward an environmentally aware citizenry that has the knowledge, skills, and moral values to protect and preserve the environment and to achieve sustainable development.

The first two premises recognize the essential complementarity of human responsibility with what is best described as awe and wonder, responses that are more easily described in poetic or religious language than in the words of scientists or lawyers. They focus on an attitude to nature that is neither a claim to rights (either human or nature's) nor—recalling Lynne White's thesis (p. 111)—an assertion of domination. Premises 3 and 4 are also complementary, asserting the role of individuals and states, respectively. In addition, premise 3 recognizes the existence and importance of the global commons and of "transgenerational equity."[3] Premise 4 lays an additional charge on states to develop renewable energy.

Premise 5 is a commitment to eradicate poverty. This is a key and non-negotiable condition for developing nations, but this premise puts their demand firmly in the context of sustainability and demography, which are often ignored. Economists tend to express this principle in terms of maintaining capital, regarding it legitimate to substitute a declining asset by one more readily available (such as plastic for metal, wind or wave energy for fossil fuel).

Premises 6, 7, and 8 are three widely accepted and almost as widely ignored principles: the precautionary principle, the "polluter pays" principle, and the need for environmental impact assessment. The last is a necessary corollary of the precautionary principle, but it warrants inclusion, not least because it implies monitoring (premise 9), which is especially important in view of the slow and often insidious nature of environmental change. Premise 10 returns to the necessary partnership between the state and its citizens, which is an essential part of any functional code of conduct.

The ten premises bring us firmly back to the notion of stewardship. They describe responsibilities rather than rights. There are both practical and theoretical reasons for this: responsibility involves response from those able to influence actions, whereas a right is merely a status. Responsibility is implicit in the relationship between states, people, other living beings, and the Earth and its environment as a whole.

In contrast, rights language implies a static relationship between rights-givers and rights-claimers, whereas the relationship between humankind and environment is a dynamic one. Rights are a lawyer's delight and ought to be anathema to any environmentalist concerned with responsible care and protection of the natural world. Moreover, they are prejudicial to developing a common approach, because they erect barriers by establishing categories for protection rather than concentrating on the processes that determine the categories.

The ten premises give us some apparently universal themes to work with. Furthermore, they link what strike mind and heart as sound principles with the fear of what may happen if we ignore them. They challenge us to find policies for long-term sustainability, providing justice to rich and poor people and nations alike. Our resources are finite. We only have one world. It would be foolish to think that environmental care will be achieved solely by ethical pronouncements or declarations from international meetings. But the last few decades have made us far more aware of our role on the Earth as a keystone species. Hopefully the ten premises might help to change the way we think *and* act.

CHAPTER 8

God's Two Books

THE MONASTERY at Assisi, Italy, a great complex of basilicas and castles on the western slope of a mountain, is part of a World Heritage Site. In 1986 the World Wide Fund for Nature (WWF) held its twenty-fifth anniversary in Assisi. The meeting culminated in receiving "Messages on Man and Nature" from the major world faiths. The Duke of Edinburgh, the fund's president, challenged the gathering:

> It is not enough just to be concerned about the conservation of nature, neither is it enough to have the scientific expertise to enable us to achieve the conservation of nature. We also need a clear and sufficient motive to ensure that our hearts as well as our minds are committed to the cause. We need the knowledge plus commitment. We need a credible philosophy. What we need is to establish the practical and moral reasons why conservation is important and to clarify the motives that will help people to commit themselves to the cause of conservation. . . . There can be little advantage in attempting to save our souls or to seek enlightenment or salvation if our very existence in this earth is threatened by our own destructive activities.

Assisi, of course, is the birthplace of St. Francis, revered for centuries for his love of nature and animals, whom, we are told, he called his brothers and sisters. When we think of St. Francis, an attitude of humility and simplicity toward nature strikes us, rather than any sort of science.

Does this attitude connect with the relationship between humanity and the natural world as seen by the Duke of Edinburgh? At one level, we are simply animals, but it is naïve to think of ourselves as only animals. A harsh critic of such reductionism has been the economist and Noble Laureate, Amartya Sen. He has challenged us to reexamine our humanness and accept that we are much more than merely consumers. He points out that we have values as well as needs and that we cherish our ability to reason, act, and participate. If the core meaning of sustainability involves acting "without compromising the ability of future generations," we must preserve—and even expand—our existing freedoms. Sen argues that a practice of "sustainable consumption and production" need not lead to any sacrifice in living standards. Sen's prompt that we are more than survival machines has strong religious implications.

We saw earlier that the historian Lynn White, in his criticism of the Christian belief in "dominion" as leading to the environmental crisis, recommended that the attitude of St. Francis be adopted as an antidote. Others have suggested that another Italian saint, Benedict, who lived seven hundred years before Francis, is a better model. Benedict encouraged his followers to care for the natural world in practical ways. In his "rule," he laid down principles of working with the land and the seasons, and developing responsible technology. He can be regarded as a supporter of an idea that became important in later centuries, that God wrote two books—a book of words (the Bible) and a book of works (creation). The two books are both written by God, but in very different languages. The challenge is to read these different languages to find where the Bible, for example, fits with the scientific knowledge of evolution, or where theological ideas such as salvation relate to our role on Earth as stewards of nature.

To explore this, we need to go back to the time of Charles Darwin and look at Darwin's own apparent acceptance of the two books. It is relevant, too, to review how other Christian leaders have faced the challenge. We will also look at specific elements of the Bible that relate to ecology and the human role on Earth.

SCIENCE AND BELIEF IN DARWIN'S DAY

Charles Darwin was born in Shrewsbury on February 12, 1809 (the same day as Abraham Lincoln). He followed his father and elder brother to medical school in Edinburgh (1825–1827), but found himself too squeamish for a medical career. He transferred to Cambridge University, reading for a general degree (1828–1831), and followed this by five more challenging and formative years (1832–1836) as a "gentleman naturalist" on HMS *Beagle*, commissioned under the command of Robert Fitzroy to survey the southern coasts of South America. Darwin's assumption had been that he would seek ordination after his Cambridge degree. He wrote home at an early stage of the *Beagle*'s voyage, "Although I like this knocking about, I find I steadily have a distant prospect of a very quiet parsonage & I can see it even through a grove of Palms."

After leaving Edinburgh, he had read with approval the evangelical bishop of Chester, John Bird Summer's *Evidences of Christianity*. At Cambridge, he was required to study William Paley's *Evidences of Christianity*. He found Paley's logic "irresistible." In his *Autobiography* he notes, "The logic of this book and I may add of his *Natural Theology* gave me as much delight as did Euclid."

During his time on the *Beagle*, Darwin began to drift away from the idea of a career as a clergyman. He did not become an atheist; in his *Autobiography* he insisted that he always continued to believe in some form of God. Near the end of his life, he wrote to the atheist John Fordyce, "It seems to me absurd to doubt that a man may be an ardent Theist and an evolutionist.... As to my own views...I have never been an atheist in the sense of denying the existence of a God."

Nevertheless, as Janet Browne comments, "It is clear that his kind of belief, though orthodox, was a very loose, English-style orthodoxy in which it was far less trouble to believe than it was to disbelieve.... For Darwin, as for countless others, belonging to the Church of England was as much a statement of social position and attitude than it was a profession of any particular doctrine.... No sane man could believe in miracles, he decided.... Yet he went to church regularly throughout the voyage, attending the shipboard ceremonies conducted by Fitzroy and services on shore whenever possible." Many years later Darwin commented, "My theology is a simple muddle. I cannot look at the universe as a result of blind chance, yet I can see no evidence of beneficent design, or indeed a design of any kind in the details."

Darwin lived in a world in a state of flux. The pervading influence was the Enlightenment, which set the tone for the succeeding Victorian decades. But assumptions about the status of humankind were changing significantly. We have seen that as early as 1691 John Ray wrote in *The Wisdom of God Manifested in the Works of Creation*, "It is a generally received opinion that all this visible world was created for Man . . . yet wise men nowadays think otherwise" (p. 71). Reviewing the post-Enlightenment period, the historian Keith Thomas judged that "at the start of the early modern period, man's ascendancy over the natural world was the unquestioned object of human endeavour. By 1800 it was still the aim of most people . . . but by this time the objective was no longer unquestioned." The implications of this were uncomfortable. Disenchanted by some of his encounters on the *Beagle*, particularly among the natives of Tierra del Fuego, Darwin wrote, "Man in his arrogance thinks himself a great work worthy of the interposition of a deity. More humble & I think truer to consider him created from animals."

The uncertainties of the time were well summarized by Don Cupitt, a modern Cambridge University clergyman:

> Mechanistic science was allowed to explain the structure and workings of physical nature without restriction. But who designed this beautiful world-machine and set

it going in the first place? Only Scripture could answer that question. So science dealt with the everyday tick-tock of the cosmic framework and religion dealt with the ultimates: first beginnings and last ends, God and the soul. . . . It was a happy compromise while it lasted. Science promoted the cause of religion by showing the beautiful workmanship of the world. . . . But there was a fatal flaw in the synthesis. Religious ideas were being used to plug the gaps in scientific theory. Science could not yet explain how animals and plants had originated and had become so wonderfully adapted to their environment—so that was handed over to religion. People still made a sharp soul-body distinction and the soul fell beyond the scope of science—so everything to do with human inwardness and personal and social behavior remained the province of the preacher and moralist.[1]

This was the world in which Darwin lived and worked. His theory gave an entirely naturalistic explanation of how all life, including human beings, emerged. His 1859 book, the *Origin of Species*, showed in a understandable way how life came into being without the need to assume any supernatural agency or direct intervention by God. The reactions of his contemporaries were by no means all negative. Charles Kingsley, author of the *Water Babies* and professor of modern history at Cambridge University, welcomed the *Origin of Species*: "I find it just as noble a conception of Deity to believe that He created primal forms capable of self development . . . as to believe that He required a fresh act of intervention to supply the *lacunas* [gaps] which He himself had made."

One person who vehemently opposed the book on both scientific and religious grounds was Adam Sedgwick, Darwin's teacher and professor of geology at Cambridge University. He reviewed *Origin of Species* in the *Spectator*:

I have read Darwin's book. It is clever and calmly written; and therefore the more mischievous if its principles be false; and I believe them *utterly false*. . . . It repudiates all reasoning from final causes; and seems to shut the door upon any view (however feeble) of the God of Nature as manifested in His works. From first to last it is a dish of rank materialism cleverly cooked and served up. . . . And why is this done? For no other solid reason, I am sure, except to make us independent of a Creator.

Another perspective came from Darwin's ally, the natural scientist Thomas Henry Huxley. He wrote, "Exhausted theologians lie about the cradle of every science as the strangled snakes beside that of Hercules; and history records that whereas science and orthodoxy have been fairly opposed, the latter has been forced to retire from the lists, bleeding and crushed if not annihilated; scotched if not slain." (A modern parallel might be found in the words of Oxford zoologist Richard Dawkins: "Although atheism might have been tenable before Darwin, Darwin made it possible to be an intellectually fulfilled atheist.")

Notwithstanding, many of the more orthodox theologians of Darwin's day took a similar approach to Kingsley, believing that evolution helped theology because it provided a mechanism for providence, a way in which God might be working out his purposes step by step, in both history and nature. A few decades after the *Origin of Species* appeared, the Oxford theologian Aubrey Moore wrote that evolution did the work of a friend under the guise of a foe. It supported the idea of an active Creator working through evolution rather than the absentee landlord implied by Enlightenment deism. For Moore, Darwinism was

infinitely more Christian than the theory of "special creation" for it implies the immanence of God in nature, and the omnipresence of His creative power. . . . Deism, even when it struggled to be orthodox, constantly spoke of God as we might speak of an absentee landlord, who cares nothing for his property so long as he gets his rent. Yet nothing more opposed to the language of the Bible and the Fathers can hardly be imagined. . . . For Christians the *facts of nature* are *the acts of God*. Religion relates these facts to God as their Author, science relates them to one another as integral parts of a visible order. Religion does not tell us of their interrelations, science cannot speak of their relation to God. Yet the religious view of the world is infinitely deepened and enriched when we not only recognize it as the work of God but are able to trace the relation of part to part.

Two years after Darwin died in 1882, Frederick Temple, the bishop of Exeter (and soon to become archbishop of Canterbury), gave the theory of evolution an imprimatur in his Bampton Lectures: "[God] did not make the things, we may say; but He made them make themselves."

The assimilation of Darwin's ideas by conservative theologians was also taking place in North America, where more extreme attitudes to biblical fundamentalism than in England were common. Between 1910 and 1915 a series of booklets called the *Fundamentals* was published in the United States to expound the "fundamental beliefs" of Protestant theology, as defined by the General Assembly of the American Presbyterian Church. Some of the booklets that addressed science were sympathetic to evolution. Princeton University theologian B. B. Warfield, a passionate advocate of Bible inerrancy, wrote the *Fundamentals'* chapter on "The Deity of Christ," declaring, "I do not think that there is any general statement in the Bible . . . that need be opposed to evolution." Warfield believed that evolution could provide a tenable "theory of the method of divine providence in the creation of mankind."

A recurring problem in debates about Darwinism is that different people have different agendas. Huxley, Darwin's "bulldog," was concerned about freedom for science from what he regarded as illegitimate pressures from theologians (and others). Although he called himself an "agnostic" about religion (he actually invented the word *agnostic*), Huxley was certainly not against religious ideas. Late in life he wrote,

> It is the secret of the superiority of the best theological teachers to the majority of their opponents that they substantially recognize the realities of things, however strange the forms in which they clothe their conceptions. The doctrines of predestination; of original sin; of the innate depravity of man and the evil fate of the greater part of the human race; of the primacy of Satan in this world, faulty as they are, appear to me to be vastly nearer the truth than the "liberal" popular illusions that babes are all born good and that the example of a corrupt society is responsible for their failure to remain so; that it is given to everybody to reach the ethical ideal if he will only try.

"SURVIVAL OF THE FITTEST" AND CHANCE

So why do some people, including many Christians, still maintain, often vehemently, that Darwin's theory is the direct cause of atheism and disbelief? Certainly one factor is that the publication of *Origin of Species* coincided with a time when religious belief was declining through the rationalism of the post-Enlightenment age of Darwin. It was easy to assume that Darwinism precipitated unbelief. Eight years before the *Origin of Species* appeared the English poet Matthew Arnold mourned the seemingly inexorable advance of science and materialism—and with it an ebbing tide of faith:

The Sea of Faith
Was once, too, at the full, and round earth's shore
Lay like the folds of a bright girdle furled.
But now I only hear
Its melancholy, long, withdrawing roar.

Later commentators would tie Arnold's pessimism to the ferment stirred by Darwin's ideas. But in fact, Arnold's unease about the fate of religion in a scientific age has much deeper roots; Darwin, Darwinism, and evolution were merely a step—albeit a significant one—in our growing understanding of our surroundings and how we relate to them.[2]

Essentially, Darwin's theory of evolution was a scientific advance and helped us understand how we relate to the natural world. Once Darwin published his ideas, however, anybody could run with them to make whatever ideological assumptions they wished. Two of these ideological arguments ended up being troublesome for religion. The first is the idea that human morals must succumb to a ruthless "survival of the fittest," and second is the idea that human origins, and even the great wonders of the natural world, are produced merely by "chance."

The first problem—survival of the fittest—is easy to dismiss. In his own research, Darwin used the way that selection by farmers had changed domesticated breeds of animals as a model for the processes happening in nature. It was his contemporary, Herbert Spencer, a railway engineer turned philosopher, who coined the phrase "survival of the fittest" to describe the process; Darwin himself did not like it. The phrase has been a source of much confusion and has led to repeated accusations that natural selection is tautologous. This is wrong: "fitness" in its biological sense refers to reproductive success, not health or physical prowess. The fittest are simply those with a characteristic that enables them to raise more offspring than the less fit—which properly describes natural selection. But the real problem was that Spencer extended Darwinian selection into sociology; he saw human history as the strong overcoming the weak in the industrial and financial spheres, a kind of ruthless progress toward more superior humans. What Darwin spoke of as biological fitness was *reproductive success*, not health or physical prowess. This concept is no threat or implication to morals.

Can Chance Make an Eye?

We can no more accept the principle of arbitrary and casual variation and natural selection as a sufficient account of the past and present organic world than we can receive Laputan method of composing books [i.e., emanating from the fictional nation described in *Gulliver's Travels*] as a sufficient one of Shakespeare and the Principia.
—JOHN HERSCHEL, REVIEWING THE *ORIGIN OF SPECIES*

A more subtle challenge to theology and morals is the belief that Darwinian evolution depends on mere "chance" and is an entirely negative influence incapable of producing seemingly perfect adaptations like a mammalian eye or the pattern of a butterfly's wing. We need to deal with this at greater length. Darwin was well aware that the idea would come under attack. In the *Origin of Species*, he devoted an entire chapter to "difficulties" to his theory and he wrote another chapter on "objections."

A common criticism of natural selection is that it depends on mutations that arise at random. This is a frequent accusation of "creationists," but extends much wider than the beliefs of anti-evolutionists. The criticism, however, is misplaced. The Darwinian process does not depend on chance; adaptation results from the selection of advantageous variants and this is a deterministic process. On the other hand, the *origin* of inherited variation is random, depending on chemical mistakes as the DNA helix divides and replicates (mistakes that we call "mutation"), and, much more significantly in sexual organisms, on the phenotypic expression of new variants resulting from the recombination of already existing variants on the chromosomes handed on to us by our parents. Any confusion arises through conflating these two processes. "Darwinian evolution" itself is not the result of chance.

What about complex organs like the eye? An important point to note here is that the function of any organ may change significantly during evolution. In the *Origin of Species*, Darwin quoted Agassiz's work on starfish, showing how spines may be modified, leading to the development of an apparently new and important trait: tube feet. Many such examples of changed use are known; they are open to investigation and test. As far as the eye is concerned, some simple organisms have no image-forming eyes, only light-sensitive cells. Any inherited variants that allow detection of the direction of light and its intensity would give advantage to their possessor, and hence be subject to natural selection.

The credibility of such a process has grown enormously since genome mapping became a reality, revealing a remarkable and previously unsuspected developmental flexibility. For example, the genes responsible for the crystalline proteins that make up the eye lens have repeatedly "redeployed" since their origin as producers of stress-related proteins in microorganisms. The Cambridge paleontologist Simon Conway Morris has outlined a host of similar cases of seemingly random features being brought into good use by the working of natural selection. In his book *Life's Solution: Inevitable Humans in a Lonely Universe*, he lists "20 or even more independent lines of differentiation [toward eye perfection], including at least 15 cases of independent attainments of photoreceptors with a distinct lens."

Furthermore, genes are quite frequently duplicated. These "spare genes" are then available to the organism to be used for new kinds of functions. While it is true that adaptation relates to survival and the possibility of gene transmission rather than long-term purpose, it is wrong to claim that "Darwinian evolution" is an entirely fortuitous process. An appearance of progress may result from the fact that, as Simon Conway Morris has pointed out, the range of viable options for any trait is very limited. Natural selection is the only mechanism (apart from direct divine intervention) which fits organisms to their environment.

The Nature of Genetics

A problem that troubled Darwin himself was that variation is apparently lost in every mating, because offspring are, generally speaking, intermediate between their parents. This was resolved after Darwin's time by the recognition that the inherited elements (genes) are transmitted unchanged between generations. The appearance of blending is because the expression of every gene is modified by other genes. The existence of particulate inheritance was Gregor Mendel's finding, published in 1865 but only realized as significant when it was rediscovered in 1900. But in solving one problem, it raised another for the Darwinians: the genes studied by the early geneticists (or Mendelians, as they were called) were almost all deleterious to their carriers, had large effects, and were inherited as recessives—all properties that seemed counter to the progressive gradualism expected under Darwinism.

This impasse persisted and widened through the first decades of the twentieth century. There were no real doubts that large-scale evolution had occurred, but it did not seem to have been driven by natural selection. Stanford entomologist Vernon Kellogg spoke of "the death-bed of Darwinism" in his introduction to a book written in 1907 for the Jubilee of *Origin of Species*. He wrote, "Darwinism as the all-sufficient or even the most important causo-mechanical factor in species-forming and hence as the sufficient explanation of descent, is discredited and cast down."

Genetics and Darwinian evolution seemed to contradict each other; biologists and geneticists were at loggerheads. Many scientists followed Kellogg and declared that Darwinism—with natural selection at its heart—had failed An extravagance of evolutionary theories poured into the void left by Darwinism's discrediting:

Leo S. Berg's *Nomogenesis*, J. C. Willis's *Age and Area*, Jan Smuts's *Holism*, Hans Driesch's entelechy, and Henry F. Osborn's aristogenesis and orthogenesis. Invention was rife. A common feature was some form of inner progressionist urge or *élan vital*. The idea that evolution was driven by some sort of purpose was influentially espoused by some distinguished scientists—the zoologist Ray Lankester and physiologist J. S. Haldane; the psychologists Lloyd Morgan, William McDougall, and E. S. Russell; physicists like Oliver Lodge; and cosmologists such as Arthur Eddington and James Jeans; as well as by popularizers like Arthur Thomson and politicians such as Arthur Balfour. Without such a force, they asked, how could deleterious mutations be overcome? Unfortunately in view of future developments, three standard and still-read histories of biology (by Erik Nordenskiöld, Emanuel Rádl, and Charles Singer) were written during this time, perpetuating the idea that evolutionary theory was an illogical mess and that Darwinism was completely eclipsed.

Not surprisingly with such apparently informed support, these ideas were seized upon by churchmen, prominent among them being Charles Gore, and somewhat later W. R. Inge, Hensley Henson, Charles Raven, and Edward Barnes.

The theological urge to find progressive forces inside nature proved wholly unsuccessful. However, it died largely because it was contradicted by outside developments, rather than being consciously rejected or disproved by science. The Modernists, who had eagerly embraced the idea of progressive forces at work, were marginalized when events such as the world wars showed the reality of human nature as cruel and unforgiving, destroying the belief of an ever-upward human progress. In the churches, there was a reawakening of the old doctrines of human sinfulness and alienation from God. One can have some sympathy with the progressive theologians who had jumped on the bandwagon of entelechy and *élan vital*. It took the scientists several decades to resolve the apparent conflict between genetics and natural selection. Nevertheless, the modernist theologians cannot be wholly excused for being so uncritical in how they used tentative science.

The irrelevance of the frenzy of evolutionary speculating in the early days of genetics was exposed in a series of theoretical analyses in the 1920s, beginning with two difficult and largely non-understood papers by the British mathematician and geneticist R. A. Fisher, a practicing Christian. He showed that continuous variation could arise through the cumulative effect of many genes, each with a small effect, and that dominance was the result of interaction between genes rather than an intrinsic property of a gene by itself. Fisher argued that the dominance or recessivity of any character is the consequence of a gene repeatedly mutating during evolutionary time. Since its expression is being modified by other genes, there will be selection for greater expression if the character is beneficial to its carrier or lesser if it has deleterious effects. In other words, the inheritance of the character would be modified toward dominance or recessivity respectively, thus removing a major difficulty raised by the early Mendelians. In this way, Fisher restored the plausibility of genetic gradualism and the idea of helpful mutations, two keystones of Darwin's theory.

Fisher's analyses were complemented by J. B. S. Haldane in Britain and Sewall Wright in the United States, and summarized by them in a series of major works. They were supported by ecological studies of inherited variation in natural populations by E. B. Ford (1931) in Britain and Theodosius Dobzhansky (1937) in the United States. These conclusions together with results from many other sources were brought together by Julian Huxley in *Evolution: The Modern Synthesis* (1942), which provided the eponymous name for the incorporation of Mendelian genetics into the insights of Darwin, and the final reconciliation of the earlier evolutionary debates.

The modern synthesis (commonly called the neo-Darwinian synthesis) of the 1940s has proved to be a robust understanding of evolutionary processes as well as a justification of Darwin's original theory. Its most serious challenge came in the 1960s, when biochemical techniques of protein analysis were applied to variation in natural populations and unexpectedly large amounts of inherited variation discovered—the large amounts of heterozygosity already described (p. 89). Conventional understanding was that such high levels of variation would lead to an unsupportable "genetic load" because the less advantageous allele would reduce the reproductive potential of its carrier. The simplest escape from this dilemma was to assume that such biochemical variants had no effect on their bearer, that is, that they were neutral and thus not subject to selection. This argument seemed to be supported by apparently regular rates of accumulation of new variants (mutations), to the extent that a molecular (or protein) clock could be calibrated on the basis of the number of gene differences between two lineages.

However, it soon became clear that the protein clock kept very poor time. Different proteins in the same organism could change at rates differing by two orders of magnitude, and even the same protein may change faster (or slower) in different groups. These findings are exactly what would be expected if the proteins were subject to selection rather than a physically determined mutation rate.

The falsifying of extreme neutralist assumptions led to attention being refocused on environmental (or ecological) factors in evolution. For example, it is improper to speak of *the* selective effect of an inherited character. Selection varies in both time and space. It may be density or frequency dependent or independent (see Figure 2.11 on p. 67). In the well-known example of moths being selectively predated by birds, the chance of being eaten (i.e., the intensity of selection) varies with the amount and history of atmospheric pollution. Places where black moths had a high survival rate have changed as pollution has declined following clean air legislation, and the survival rate of black moths has declined in proportion (see Figure 2.1 on p. 32).

One positive outcome of this debate has been to rescue evolutionary studies from the danger of overdependence on theoretical models and lead us back to an observational and experimental basis—which is where Darwin and Wallace began.

HUMAN EVOLUTION

The biggest theological challenge for Darwinian ideas, however, is not the fact or mechanism of evolution *per se*, but human evolution. Anxious to minimize controversy, Darwin steered clear of human evolution in the *Origin of Species*. He included only one mention of the subject: "I see open fields for far more important researches. . . . Much light will be thrown on the origin of man and his history." But he could not ignore the topic altogether and returned to it fourteen years later with *The Descent of Man*. He expected "universal disapprobation, if not execution"—meaning, he told a critic, that the book would "quite kill me in your good estimation."

Darwin knew almost nothing about human fossils. The first Neanderthal fossils were only described in 1856. But, as we have seen, subsequent discoveries have led to *Homo* having a more complete fossil record than almost any other genus, and the findings of molecular biology have shown the remarkable similarities between human and ape genomes. For some, these data strengthen the assumption that humans are nothing but apes. This conclusion is neither universal nor inevitable.

The Bible or a "Naked Ape"

In the 1960s a well-written book called *The Naked Ape* by British zoologist Desmond Morris caught the public imagination. It popularized the field of human ethology—the study of human behavior as animal behavior. Morris told a story radically different from the account of humankind found in virtually all religions, particularly the account in Genesis, an authoritative document for Christianity, Judaism, and Islam. *The Naked Ape*, of course, is only one such purveyor of such skepticism. Other notorious tracts are Richard

Dawkins's *The Blind Watchmaker*, Daniel Dennett's *Darwin's Dangerous Idea*, and Sam Harris's *Letter to a Christian Nation*. What should we make of the notion of human beings as "naked apes"?

The genetic history of mankind—the origins of our physical form—is not the real problem. The challenge we face is our ability, emotionally and intellectually, to ask how (or if) evolutionary science meshes with our humanness. We can, of course, simply deny that we are uniquely distinct from the apes in any significant way. This view is entirely possible, but it is based merely on a resolute faith that we are naked apes only. What happens if we acknowledge our distinctiveness and put together all our understandings, theological and scientific?

Humans are obviously different from the apes in many ways, but are the differences merely ones of degree or is there a real qualitative difference? This question is probably unanswerable from a scientific point of view, but theologically there is a simple solution: to regard the biological species *Homo sapiens*, descended from a primitive simian stock and related to living apes, as having been transformed by God at some time in history into *Homo divinus*, biologically unchanged but spiritually distinct. The Bible tells us that we are made of the dust of the ground, like all other living things. But it tells us also that we are made "in the image and likeness" of the Creator. This *imago Dei*, God's image in us, need not leave any anatomical or genetic traces. God's image is certainly not physical. To bring the idea of *Homo divinus* into the scientific discussion (about *H. sapiens*) we have to introduce a religious assumption: that the Creator, by a divine process, has endowed humanity with uniqueness. It is an assumption also at the heart of Genesis.

Was God's image imparted suddenly, or did it emerge gradually? Does the distinction really matter? Genesis is explicit that humanity is created in the image of God. But it leaves open how—and when—this took place. Genesis 1 describes the appearance of *H. divinus* as a *bārā* event, a specific act of God. On the other hand,

Genesis 2:7 describes the event as a divine in-breathing into an already existing entity. We should not read either account as implying that a 'soul' was inserted into the human form. Indeed scriptural exegesis and modern neurobiology both reject the notion that the advent of *H. divinus* meant the addition of a soul to a body.[3] In the Genesis account, "the man"—humanity—from all the creatures, is addressed directly by God (Gen. 1:28). Perhaps the best way of understanding this is to say that God's grace in creating us involved establishing a relationship with him, not merely tacking something on to human existence. Humans are created in such a way that their very existence is intended to be their relationship to the Creator.

The Hebrew words used in Genesis for God's creating work give us some help. Genesis 1 uses two different Hebrew terms to describe God's creative work: *yasah* (which has the sense of modeling from previous material, as a potter molds clay) and *bārā*, which is used to refer to the creation of matter (v. 1), the great sea monsters (v. 21) and mankind (v. 26), and which is always used in the Bible to refer to God's creative activity. It would be dangerous to build too much on this use of words. Both refer to a divine work, and no clear demarcation exists in the Bible between God's works in nature and his works in history; he is sovereign in both. But the meaning seems indisputable. Mankind in its entirety is made as a creature in God's image. The creation of humankind is something far different from that of all the other living creatures. (The use of *bārā* in relation to the great sea creatures probably reflects the text's insistence that they are part of God's work as opposed to representing his opponents as in some of the ancient creation stories.)

It was truly a unique event. It must have taken place after the emergence of *Homo sapiens* and may have been quite late in time. The Bible describes Adam as a neolithic farmer. It does not seem critical to decide whether God worked through some emergent process or more suddenly, or when this activity took place. The important factor is that the creation of *H. divinus* was a divine act and therefore not susceptible to usual methods of enquiry.

From beginning to end, the Bible speaks of our interdependence with the rest of creation. Sometimes we are given direct commands, as when we are told to "have dominion," which we must be careful to interpret in the way that the ancient Israelites understood royal rule—as caring servanthood (as in Ps. 72), not as despotic control as condemned by Lynn White. In other places, God's instructions are implicit (the perils of a journey, the care needed for a farm or a flock of animals, the mastery we may expect over wild animals or fierce weather). We are told that sin led to Noah's flood (Gen. 6:5–7), but also to drought (Lev. 26; Deut. 28). Furthermore, the food laws regulated hunting. A very positive attitude to creation is expressed in the Wisdom Literature.

As we have seen, in our modern day we are increasingly learning how much we depend upon creation's (nature's) services. In all cases we interact with creation. We are a part of it as well as apart from it. The Bible tells us the same story but in another way: this is God's world and he has made a covenant with us that he has promised to uphold. In the New Testament, Paul explicitly links creation and us as individuals being reconciled with Christ's death on the cross (Col. 1:20).

Complementarity

That there is a limit upon science is made very likely by the existence of questions that science cannot answer and that no conceivable advances of science would empower it to answer. . . . I have in mind such questions as "How did everything begin?" "What are we all here for?" "What is the point of living?"
PETER MEDAWAR

The real problem for the believer about evolution is to understand how God can work in an apparently deterministic universe. On the face of it, nature might seem like a machine, with no room for God to play any role (implying that we are mere automata, with no possibility of free will), but this approach oversimplifies. A helpful way to understand how God may act in his world is the idea of complementarity, effectively recognizing that nature, our experiences, and God operate at different levels and can be explained in different ways.

The complementarity model builds on proposals developed by Niels Bohr. In his unraveling of the nature of subatomic particles, Bohr recognized that the electron behaves—apparently—both as a wave and a particle: one phenomenon with two realities, both correct. Bohr then suggested that this model could be extended to other phenomena, such as mechanistic and organic models in biology, behavioral and introspective models in psychology, models of free will and determinism in philosophy, and even models of divine justice and divine love in theology. Others have edged toward a similar idea for God's action, e.g. theologian Austin Farrer's proposals about "double agency," pioneering chemist Michael Polanyi's ideas about "levels of explanation," but for our purposes, perhaps the most robust and satisfying elaboration has come from the Scottish physicist and neuroscientist Donald MacKay (1922–1987). For MacKay,

> The God in whom the bible invites belief is no "Cosmic Mechanic." Rather is he the Cosmic Artist, the creative Upholder, without whose constant activity there would be not even chaos, but just nothing. What we call physical laws are expressions of created events that we study as the physical world. Physically they express the nature of the entities "held in being" in the pattern. Theologically they express the stability of the great Artist's creative will. Explanations in terms of scientific laws and in terms of divine activity are thus not rival answers to

the same question; yet they are not talking about different things. They are (or at any rate purport to be) complementary accounts of different aspects of the same happening, which in its full nature cannot be explained by either alone. To invoke "natural processes" is not to escape from divine activity, but only to make hypotheses about its regularity.... (For example, we cannot settle the validity of our ideas in geometry by discussing the embryological origin of the brain!)

MacKay used a painting to explain his point. We can describe a painting in terms of the distribution of chemicals on a two-dimensional surface and also as the physical expression of a design in the mind of an artist. In other words, the same material object can have two or more "causes," which do not contradict or overlap, but are undeniably complementary. MacKay then extended the idea into a dynamic form, using the analogy of a television program that can be "explained" in terms of electronics and the physiology of vision, but also as the intention of the program producer who is telling the story—and who can change the images at will. The idea that the same event can have multiple causes is not new; it is at least as old as Aristotle in his *Metaphysics*, where he distinguished material and efficient causes (which answer, in general, "how" questions) from formal and final causes (which answer "why" questions).

For MacKay and those who accept his approach, people can describe and analyze an event in as quantitative and rigorous way as possible, but also acknowledge God's hand in and control of it. Complementarity has its own rules. It has suffered because it has been sometimes used improperly to link (or explain away) contrasting explanations. But complementarity can be a powerful and satisfying way to bring together scientific and religious explanations. In the context of evolution, it is entirely logical to believe in God as creator and sustainer and simultaneously accept a conventional scientific account.

The Two Books

God writes the gospel not in the Bible alone,
but on trees and flowers and clouds and stars.
MARTIN LUTHER

In the Bible, Psalm 111:2 ("Great are the works of the Lord, studied by all those who delight in them") is sometimes called the
"Research Scientist's Text." The words were carved into the wooden
door of the old Cavendish Laboratory of Cambridge University at
the behest of the first Cavendish professor, James Clerk Maxwell.
In 1953 they were the doors through which Francis Crick and James
Watson rushed to the nearby pub "to tell everyone within hearing
that we had found the secret of life." (That day, they had decided
on their model of the DNA double helix, which was correct.) The
verse also appears on the memorial plaque to Darwin's friend and
confidant, Joseph Hooker, in St. Anne's Church, Kew.

The words recall a tradition we spoke of earlier, that God wrote
two books—a book of words (the Bible) and a book of works
(creation). Darwin himself apparently endorsed the idea when he
included a quotation from Francis Bacon's *Advancement of Learning* opposite the title page of the *Origin of Species*: "Let no one think
or maintain that they can search too far or be too well studied in
the book of God's words or in the book of God's works; rather, let
all endeavor an endless progress or proficiency in both." Properly
understood, there can be no conflict between God's revelation of
himself in words or in work. Whatever the reasons for the ebbing of
faith feared by Matthew Arnold in his *Sea of Faith* lament, assaults
by science should not be demonized as a crucial one.

The understanding of humans genetically related to the apes but
uniquely made in God's image can give a coherent understanding of how God may relate to and work in the world. It is where
believers depart from refuseniks (those who, by a defiant act of
faith, deny there is a God). Such an understanding is the starting
point for recognizing the nature of full humanness, while still leav-

ing open many theological questions. For example, why is there so much suffering and death in the world? This was a major problem for Darwin himself. He wrote to Asa Gray, professor of botany at Harvard University and also a Christian:

> I own that I cannot see, as plainly as others do & as I shd wish to do, evidence of design & beneficence on all sides of us. There seems to me too much misery in the world. I cannot persuade myself that a beneficent & omnipotent God would have designedly created the Ichneumonidæ [parasitic insects] with the express intention of their feeding within the living bodies of caterpillars, or that a cat should play with mice.

The question of suffering is a hard one, and always will be. But we can approach it both religiously and scientifically. In Genesis, God declared his creation to be "Good . . . very good." However, that was God's judgment, not ours (see also Isa. 55:8). The definition of "good" in Genesis refers to God's assessment of his achievement, not necessarily a final state of perfection. We can be confident of this, because creation was explicitly committed to us to care for and nurture it (Gen. 2:15). Crocodiles and mosquitoes are not attractive to us, but are still God's creatures.

In the Bible, the value of animals derives from the Creator, not from their usefulness to us, their market value, or their charm. For some believers, the widespread occurrence of conflict and death in the natural world mars God's work and is seen as the direct consequence of God's curse on Adam and Eve following their disobedience as recorded in Genesis 3. For others, this explanation fails as soon as it is accepted that there were millions of years of life on Earth before the appearance of mankind, and these years inevitably included death on a large scale.

Many fossils have been found of animals that lived long before humans appeared but that were diseased (such as dinosaurs with rheumatic joints). No changes in the fossil record have been detected at the time humans first appeared that can be attributed to sin. The most plausible explanation of disease is that it is an inevitable consequence of how the natural world works. Mutations are necessary to produce the wonderful variety of the world, but they may also produce disorder. Pain, while always uncomfortable, is a positive protective mechanism. If we did not feel pain, we would expose ourselves to all sorts of hazards.

More importantly, we should not automatically assume that references to death in the Bible always refer to physical death. The death from which Christ saved us is not simply spiritual insensitivity or blindness, nor is it merely a liability to physical mortality; it is a severance of relationship with God, the source of life. The New Testament tells us that "we were dead, but now in Christ we are made alive."[4] The death referred to must be something other than physical death.

In the New Testament, St. Paul argues that this "death" involves separation between nature and humanity (the creation is "groaning in travail") as well as between God and humankind. He suggests that the condemnations of Genesis 3 (the "Fall") reflect a disjointedness in creation leading to "frustration" for all living things (Rom. 8:20). As a result of this break with God, the world's discord is like an orchestra without a conductor. In other words, the consequence of the human fall is not primarily about disease and disaster, or about the dawn of self-awareness. Rather it is a way of describing a fracture in the relationship between God and the humans made in his image. The rupture means that we rattle around in our space, as it were, producing disorder within ourselves, with our neighbors, and with our environment (human and nonhuman).

The Christian answer is that this discord will continue until our relationship with God is restored and we become "at peace with God through our Lord Jesus who has given us access to the grace in which we now live" (Rom. 5:1, 2)—words that condition and explain the state of nature which Paul uses later in the same section of Romans (8: 19–21). As long as we refuse to accept our responsibility to care for creation, so long the world of nature remains frustrated and dislocated. It does not seem too far-fetched to see Adam's failure as at root a failure of stewardship, violating the very first command to the human race and hence ignoring the purpose for which we were placed on earth.

At one level, eating the forbidden fruit in the Paradise Garden was simple disobedience, but the story surely intends it to be much more. Its significance was treating God as unnecessary and irrelevant. Only when we are truly fitting into our proper place as children in relation to the Father will this dislocation be reduced. The possibility of this is, of course, the gospel: we are assured that God, through Christ, has "reconciled all things to himself, making peace through the shedding of his blood on the cross" (Col. 1:20). This is where ecology and exegesis come together and indicate that the earth's curse is not a change in ecological law, but a massive failure by a keystone species—the human species.

The Old Testament gives us repeated examples of human disobedience, many of them leading to environmental damage. Scattered around the laws given to the Israelites are specific environmental commands. Romans 8:19–22 is the most explicit New Testament passage on what God intends to do with the cosmos. Paul's argument in Romans 5–8 (indeed in the whole of Rom. 1–8) is that the renewal of God's covenant results in the renewal of God's creation. He argues the same point in 2 Corinthians 3–5.

When Paul writes in Romans, "The universe itself is to be freed from the shackles of mortality and is to enter upon the glorious liberty of the children of God," he is concluding an analogy that began with the exodus from Egypt and entry into the Promised Land. God's people come through the waters of baptism (paralleling the passage through the Red Sea), are freed from sin (slavery, in parallel with Egypt), and then given, not a text (the Torah) as in Sinai, but the Spirit.

All the Bible writers speak of a God intimately involved with his creation, not a distant impersonal Designer who has left us to get on with things on our own. Paul's message is one of hope. This hope is also conveyed as the liberation of God's people from exile, spoken of in Isaiah as the rejoicing of all creation (Isa. 55:12–13). The little book of Ecclesiastes is particularly interesting for a scientist, because the author of the book behaves exactly like a scientist. He tries various experiments—with sex and drink, with building works and with farming enterprises. None of them bring him satisfaction. He concludes, "You have heard it all. Fear God and keep his commandments; this sums up the duty of mankind." This conclusion is not unlike that of the Millennium Ecosystem Assessment and its analysis of the factors that contribute to human well-being (p. 167). It brings us full circle. If we read only one of God's books, we cannot avoid having a partial understanding of God and his plans for us—both as human beings and the world that he created and sustains.

As a young man, German astronomer Johannes Kepler (1571–1630) wrote to a friend, "I wanted to become a theologian; for a long time I was unhappy. Now, behold, God is praised by my work, even in astronomy." For him, the practice of science was "thinking God's thoughts after him." His prayer was as follows:

> If I have been enticed into brashness by the wonderful beauty of thy works, or if I have loved my own glory among men, while advancing in work destined for thy glory, gently and mercifully pardon me; and finally, deign graciously to cause that these demonstrations may lead to thy glory and to the salvation of souls, and nowhere be an obstacle to that. Amen.

Kepler's prayer is a fit and proper one for any scientist, not least an ecologist.

Acknowledgments

I AM INDEBTED to Hefin Jones, Ghillean Prance, Simon Stuart, and Robert White for reading and helpfully commenting on parts of this book in draft, but most of all to Calvin DeWitt, who read the whole text and made many positive suggestions. It goes without saying that all the remaining errors are mine. My thanks are due also to Professor Wentzel van Huyssteen and Dr. Khalil Chamcham for giving me the opportunity and challenge to write *Ecology and Environment* and bring together my academic background as an ecologist with my commitment as a Christian; to Larry Witham, the Templeton Science and Religion Series editor; Natalie Silver at Templeton Press; to Jim Bacon for drawing many of the figures; to Lynda Brooks at the Linnean Society of London; and Jane Dempster at University College London. And it is right that I place firmly on record my debt to five men who inspired me with a love and respect for the natural world and who taught me that science is not bounded by laboratories or worship by churches, chapels or mosques: Bramwell Evans, Methodist minister and BBC broadcaster; John Barrett, charismatic conservationist and field biologist; Bernard Kettlewell, physician, lepidopterist, and all-round naturalist; Charles Raven, theologian and historian; and John Stott, Christian and self-confessed practitioner of orni-theology.[1]

 Notes

PREFACE

1. Speech to the Society of Biology, delivered in London, March 25, 2010.

CHAPTER 1

1. This list comes from a British Ecological Society poll reported by Malcolm Cherrett, in a volume edited by him, *Ecological Concepts* (Oxford: Blackwell Scientific, 1989).
2. Laplace, responding to Napoleon's complaint that a book he had written on the origin of the universe did not mention the Creator, made the comment, "I have no need of that hypothesis." He was not denying the existence of the Creator, merely recording his belief in a God who worked through discoverable laws; his God was a painstaking governor, not a distant prince.
3. Andrew J. Dugmore, Christian Keller, and Thomas H. McGovern, "Norse Greenland Settlements: Reflections on Climate Change, Trade, and the Contrasting Fates of Human Settlements in the North Atlantic Islands," *Arctic Anthropology* 44 (2007): 12–36.
4. See Ian Tattersall, *Paleontology* (West Conshohocken, PA: Templeton Press, 2010), in this series, and also Ian Tattersall, *The Fossil Trail: How We Know What We Think We Know About Human Evolution*, 2nd ed. (New York: Oxford University Press, 2009).

CHAPTER 2

1. The most studied example of this is the peppered moth (*Biston betularia*), but the phenomenon occurred in over a hundred species.
2. See my book, R. J. Berry, *Inheritance and Natural History* (London: Collins, 1977).

CHAPTER 3

1. The official naturalist was Robert McCormick, who left the *Beagle* early in the voyage in pique at Darwin's position.

CHAPTER 4

1. Clarence J. Glacken, *Traces on the Rhodian Shore: Nature and Culture in Western Thought from Ancient Times to the End of the Eighteenth Century* (Berkeley: University of California Press, 1967).
2. George Perkins Marsh, *The Earth as Modified by Human Action* (New York: Scribner's, 1898).
3. Lynn White's lecture has been reprinted in many anthologies, but was originally published as Lynn White, "The Historical Roots of Our Ecologic Crisis," *Science* 155 (1967): 1203–7.
4. See a description in Arthur Lovejoy, *The Great Chain of Being* (Cambridge, MA: Harvard University Press, 1936).
5. The rationale and development of Rolston's ideas and an account of his initial rejection by both religious and secular advocates are set out in a biography of Rolston: Christopher Preston, *Saving Creation* (San Antonio, TX: Trinity University Press, 2009).
6. Lovelock describes the origin and development of Gaia in an autobiography, James Lovelock, *Homage to Gaia* (Oxford: Oxford University Press, 2000).
7. Although the Sierra Club lost the appeal, it won its war. Discouraged by long delays, Walt Disney Enterprises did not pursue its planning permission. In 1978 the U.S. Congress added the disputed territory to Sequoia National Park.

CHAPTER 5

1. The expeditions were commissioned by the Royal Society of London. The ostensible reason for the first voyage was to observe the transit of Venus across the face of the Sun to improve the determination of latitude and thus navigation; the second voyage was to search for the "Great Southern Continent."
2. Tansley's work was published in 1951 when he was eighty years old. It is reprinted in R. J. Berry and John H. Crothers, eds., *Nature, Natural History and Ecology* (London: Academic Press, 1987).

CHAPTER 6

1. See the volume in this series by Malcolm Jeeves and Warren Brown, *Neuroscience, Psychology, and Religion* (West Conshohocken, PA: Templeton Press, 2009).
2. Neander's Valley (*thal* in German) was named after Joachim Neander (1650–1680), a German theologian and hymn-writer, author of "Praise to the Lord, the Almighty, the King of Creation," based on Psalms 103, 150. He lived nearby and used to frequent the valley to contemplate the natural world and to preach.
3. Haldane first put forward this idea in a book published in 1932, but presented it in a more popular form in an article in 1955: "Let us suppose that you carry a rare gene which affects your behaviour so that you jump into a flooded river and save a child, but you have one chance in ten of being killed, while I do not possess the gene and stand on the bank and watch the child drown. If the child is your own child or your brother or sister, there is an even chance that the child

will also have this gene, so five genes will be saved in children for one lost in an adult. If you save a grandchild or nephew, the advantage is only two and a half to one. If you only save a first cousin, the effect is very slight. If you try to save your first cousin once removed the population is more likely to lose this valuable gene than to gain it. . . . It is clear that genes making for conduct of this kind would only have a chance of spreading in rather small populations where most of the children were fairly near relatives of the man who risked his life" (*New Biology* 18 [1955]: 44).

CHAPTER 7

1. Stephen Jay Gould, *Eight Little Piggies: Reflections in Natural History* (New York: W. W. Norton, 1993). Gould made this comment in the chapter, "The Golden Rule: A Proper Scale for Our Environmental Crisis," 48.
2. The seventh of the ten Millennium Development Goals is to "ensure environmental sustainability." It involves four main and ten subsidiary targets:
   1. Integrate the principles of sustainable development into country policies and programs; reverse the loss of environmental resources
   2. Reduce biodiversity loss by making a significant reduction in the rate of loss of:
      - Proportion of land area covered by forest
      - CO_2 emissions, total, per capita, and per \$1 GDP
      - Usage of ozone-depleting substances
      - Proportion of fish stocks to within safe biological limits
      - Proportion of total water resources used
      - Proportion of terrestrial and marine areas protected
      - Proportion of species threatened with extinction
   3. Reduce by half the proportion of people without sustainable access to safe drinking water and basic sanitation, specifically to increase
      - Proportion of population using an improved drinking water source
      - Proportion of population using an improved sanitation facility
   4. Achieve significant improvement in lives of at least 100 million slum dwellers by 2020 as measured by the proportion of the urban population living in slums
3. John Ruskin wrote in his essay, "The Lamp of Memory," (published in *Seven Lamps of Architecture*, 1849), "God has lent us the earth for our life; it is a great entail. It belongs as much to those who are to come after us and whose names are already written in the book of creation, as to us; and we have no right, by anything we do or neglect, to involve them in unnecessary penalties, or deprive them of benefits which it was in our power to bequeath." Former British prime minister Margaret Thatcher expressed the same idea: "We do not hold a freehold on the world, only a full repairing lease." Similar thoughts have come from many others. John James Audubon wrote, "A true conservationist is a man who knows that the world is not given by his fathers but borrowed from his children"; Theodore Roosevelt, "I recognize the right and duty of this generation to develop and use our natural resources, but I do not recognize the right to waste them or to rob by wasteful use the generations that come after us";

Dietrich Bonhoeffer, "The ultimate test of a moral society is the kind of world that it leaves to its children"; and Pope John Paul II, "The future starts today, not tomorrow."

CHAPTER 8

1. Don Cupitt, *Sea of Faith* (London: British Broadcasting Corporation, 1984), 59.
2. The Matthew Arnold poem was published in 1867, after the publication of the *Origin of Species*, but literary analysts agree that it was composed around fifteen years earlier.
3. See the volume by Malcolm Jeeves and Warren Brown in this series, *Neuroscience, Psychology, and Religion* (West Conshohocken, PA: Templeton Press, 2009). See also Joel Green and Stuart Palmer, eds., *In Search of the Soul* (Downers Grove, IL: InterVarsity Press, 2005).
4. For example, Rom, 8:10; Eph. 2:1–5; see also John 3:4–7.

ACKNOWLEDGMENTS

1. John R. W. Stott, *The Birds, Our Teachers* (Grand Rapids, MI: Baker Books, 2008).

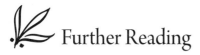 Further Reading

GENERAL TEXTS

Beeby, Alan, and Anne-Maria Brennan. *First Ecology: Ecological Principles and Environmental Issues.* 3rd ed. Oxford: Oxford University Press, 2008.

Begon, Michael, Colin R. Townsend, and John L. Harper. *Ecology: From Individuals to Ecosystems.* 4th ed. Oxford: Blackwell, 2006.

Chivian, Eric, and Aaron Bernsten. *Sustaining Life: How Human Health Depends on Biodiversity.* Oxford: Oxford University Press, 2008.

Danchin, Étienne , Luc-Alain Giraldeau, and Frank Cézilly. *Behavioural Ecology.* Oxford: Oxford University Press, 2008.

Krebs, Charles J. *Ecology: The Experimental Analysis of Distribution and Abundance.* 6th ed. San Francisco: Pearson Benjamin Cummings, 2009.

Smith, Thomas M., and Robert L. Smith. *Elements of Ecology.* 7th ed. San Francisco: Pearson Benjamin Cummings, 2009.

Southwood, T. R. E., and P. A. Henderson. *Ecological Methods.* 3rd ed. Oxford: Blackwell Scientific, 2000.

INTRODUCTORY ACCOUNTS

Botkin, Daniel B. *Discordant Harmonies: A New Ecology for the Twenty-first Century.* New York: Oxford University Press, 1990.

Colinvaux, Paul. *Why Big, Fierce Animals Are Rare.* London: G. Allen and Unwin, 1979.

Diamond, Jared M. *Collapse.* New York: Viking Books, 2005.

———. *Guns, Germs and Steel.* New York: W.W. Norton, 1997.

———. *The Rise and Fall of the Third Chimpanzee.* New York: Vintage, 1992.

Golley, Frank B. *A Primer for Environmental Literacy.* New Haven, CT: Yale University Press, 1998.

McNeill, John R. *Something New Under the Sun: An Environmental History of the Twentieth-Century World.* New York: W. W. Norton, 2000.

Nowak, Martin and Roger Highfield *Super Cooperators: Evolution, Altruism, and Why We Need Each Other to Succeed.* New York: Free Press, 2011.

Quammen, David. *The Song of the Dodo: Island Biogeography in an Age of Extinctions*. New York: Scribner, 1996.

Weiner, Jonathan. *The Beak of the Finch: A Story of Evolution in Our Time*. New York: Alfred Knopf, 1994.

HISTORICAL AND CLASSICAL BOOKS

Allee, Warder C. *Animal Aggregations: A Study in General Sociology*. Chicago: University of Chicago Press, 1931.

Carson, Rachel. *Silent Spring*. Boston: Houghton Mifflin, 1962.

Clements, Frederic E. *Plant Succession*. Washington, D.C.: Carnegie Institution of Washington, 1916.

Elton, Charles S. *Animal Ecology*. Chicago: University of Chicago Press, 2001 [1927].

———. *The Ecology of Invasions by Animals and Plants*. Chicago: University of Chicago Press, 2000 [1958].

Hagen, Joel B. *An Entangled Bank: The Origins of Ecosystem Ecology*. New Brunswick, NJ: Rutgers University Press, 1992.

Harman, Oren. *The Price of Altruism: George Price and the Search for the Origins of Kindness*. New York: W. W. Norton, 2010.

Mabey, Richard. *Gilbert White: A Biography of the Author of* The Natural History of Selborne. London: Century Hutchinson, 1986.

Marsh, George. *Man and Nature*. Cambridge, MA: Belknap Press of Harvard University Press, 1965 (1864).

McIntosh, Robert P. *The Background of Ecology: Concept and Theory*. Cambridge: Cambridge University Press, 1985.

Mitman, Gregg A. *The State of Nature: Ecology, Community, and American Social Thought, 1900–1950*. Chicago: University of Chicago Press, 1992.

Odum, Eugene P. *Fundamentals of Ecology*. Philadelphia: Saunders, 1953.

Real, Leslie A., and James H. Brown, eds. *Foundations of Ecology: Classic Papers with Commentaries*. Chicago: University of Chicago Press, 1991.

Shelford, Victor E. *Animal Communities in Temperate America as Illustrated in the Chicago Region*. Chicago: Bulletin of the Geographical Society of Chicago, 1913.

Slack, Nancy G. *G. Evelyn Hutchinson and the Invention of Modern Ecology*. New Haven, CT: Yale University Press, 2011.

HUMAN IMPACTS

Ashby, Eric. *Reconciling Man and the Environment*. Stanford, CA: Stanford University Press and London: Oxford University Press, 1978.

IUCN, UNEP, WWF. *Caring for the Earth: A Strategy for Sustainable Living*.

Gland, Switzerland: The World Conservation Union, United Nations Environment Programme, World Wide Fund for Nature, 1991.

————. *World Conservation Strategy. Living Resource Conservation for Sustainable Development*. Gland, Switzerland: United Nations Environment Programme and World Wildlife Fund, 1980.

Meadows, Donella H., Jørgen Randers, and Dennis Meadows. *The Limits to Growth: The 30-Year Update*. White River Junction, VT: Chelsea Gree, 2004.

Pimm, Stuart L. *The World According to Pimm*. New York: McGraw-Hill, 2001.

Wilson, Edward O. *The Future of Life*. New York: Alfred Knopf, 2002.

Religion and the Environment

Bauckham, Richard, *The Bible and Ecology: Rediscovering the Community of Creation*. Waco, TX: Baylor University Press, 2010.

Berry, R. J. *God's Book of Works: The Nature and Theology of Nature*. London: T&T Clark, 2003.

————, ed. *Environmental Stewardship*. London: T&T Clark, 2006.

Bouma-Prediger, Steven. *For the Beauty of the Earth*. Grand Rapids: Baker Academic, 2001.

Glacken, Clarence J. *Traces on the Rhodian Shore*. Berkeley: University of California Press, 1967.

McGrath, Alister. *Darwinism and the Divine: Evolutionary Thought and Natural Theology*. Chichester: Wiley–Blackwell, 2011.

————. *The Reenchantment of Nature: The Denial of Religion and the Ecological Crisis*. New York: Doubleday, 2002.

Reference Works

Levin, Simon A., ed. *Encyclopedia of Biodiversity*. San Diego: Academic Press, 2001. 5 vols.

————. *The Princeton Guide to Ecology*. Princeton, NJ: Princeton University Press, 2009.

Millennium Ecosystem Assessment. *Our Human Planet: Summary for Decision-Makers*. Washington, D.C.: Island Press, 2005.

United Nations Environment Programme. *Global Environment Outlook: Environment for Development, GEO 4*. London: U.N. Stationery Office, 2007.

 Index